Closeup:
New Worlds

Pioneer 11 *is launched from Kennedy Space Center, April 5, 1973. It flew past Jupiter in December, 1974 and will reach the planet Saturn in 1979.*

Closeup:
New Worlds

Edited by Ben Bova with Trudy E. Bell

St. Martin's Press • New York
St. James Press • London

All photos are courtesy of NASA unless otherwise noted.

Copyright © 1977 by Ben Bova
All rights reserved. For information, write:
St. Martin's Press, Inc., 175 Fifth Ave., New York, N.Y. 10010.
Manufactured in the United States of America

Published simultaneously in U.K. by St. James Press
For information, write:
St. James Press Ltd.
3 Percy Street
London W1P 9FA

To Robert A. Heinlein, who turned our eyes skyward.

Table of Contents

Foreword   Mark Chartrand, *Director, Hayden Planetarium*   ix

Introduction: Five Stars   Ben Bova and Trudy E. Bell   xi

Sol III—The Twin Planet   G. Harry Stine   1

The Old and New Mars   Jerry Pournelle   31

The Exploration of Venus   Gregory Benford   79

The Surprising World Called Mercury   Joe Haldeman   107

Jupiter: Eden With a Red Spot   Hal Clement   137

Styx and Stones, and Maybe Charon Too   George Harper   175

The Origin of the Solar System   Richard C. Hoagland and Ben Bova   197

…And Beyond   Ben Bova   217

Glossary   219

Foreword

DESCANT: A discourse or comment on a theme, like variations on a musical air. In *Closeup: New Worlds* the music is that of the spheres, and the theme is new knowledge of the players, new answers to questions that have been asked for thousands of years, and new questions that compel further study.

Four hundred years ago, Johannes Kepler actually wrote out the melodies he thought were played by the planets as they went around their orbits, from shrill Mercury to *basso profundo* Saturn. He tried to fit the spacing of orbits by regular polyhedra, and his mysticism was encouraged when he discovered certain mathematical relationships between the periods and sizes of the orbits.

But even as Kepler sat musing, the planets were losing their mysteries through the discoveries of an irascible Italian scholar, Galileo Galilei. Following his use of the telescope for astronomy, the first half of the 1600's saw an explosion of knowledge about our neighbor worlds, an explosion comparable only to that which has occurred in the third quarter of the twentieth century.

In 1600, *all* knowledge of the planets could be put into a few pages of a single book. Even in 1900, all studies of the solar system would have fit into a few volumes of mostly redundant information. Today many meters of library shelves, thousands of photo files, and thousands of *kilometers* of computer tape are required, and the tape is getting longer each day.

This latter-day crescendo of information is the basis of the stories here—for stories they are, even if true. You will find out what we know, and just as important, what we don't know. And unlike a textbook, the authors bring their imaginative talents to tell you what we might know, and do, in the future. As you read about our study of the planets, I hope you will appreciate the process of trial and error, guess and conjecture, hypothesis and test, false starts, and the excitement of discovery and speculation.

I hope also that you will realize why men have spent their lives studying things they may never get to: because it is fun and rewarding. If you think of scientists as humorless plodders, you don't know any, and didn't see the jubilation at the Jet Propulsion Lab when *Viking* landed on Mars. And you won't think so after reading this book.

With luck—and money—a great deal of this book will be out of date in ten years. Let's hope so, for it will mean that we will have more and better information about our corner of the fascinatingly variegated universe.

New York City
September 1976

Mark R. Chartrand III
Director, Hayden Planetarium
American Museum of Natural History

Introduction

Five Stars

The Greeks called them wanderers, *planetos*. And long before blind Homer sang of Troy, the Chaldeans and other desert starwatchers realized that five of the eternal stars of heaven refused to stay fixed in the night sky.

Obviously, these five wandering stars, or planets, were the abodes of gods. Each civilization named them after their own deities. Our civilization is a linear descendent of the Romans', and we know them by their Latin names: Mercury, Venus, Mars, Jupiter and Saturn.

Thousand of years before the birth of Christ, desert astronomers tried to puzzle out why these five wanderers can move across the firmament of fixed stars. All the other stars remained steadfastly in place, rising and setting in perfect order night after night, year after year, generation after generation. The five wanderers moved freely through the skies.

But not so freely after all, the astronomers learned. The planets' paths were restricted, restrained to certain avenues across the sky. Their motions could even be predicted, with fair accuracy.

The Greeks, who prized order above all things, worked out a logical system to explain and predict the movements of these wandering sky-gods. Eudoxus of Knidus, in the fourth century before Christ, is credited with originating the concept that the planets move on crystal spheres that revolve in perfect circles around our fixed and immovable Earth. Nearly four centuries later, Ptolemy of Alexandria wrote a book about Hipparchus' system, which survived the ravages of Europe's Dark Age, thanks to the Arab civilization that preserved much of the ancient writings. We know the book today by a corruption of its Arabic name,

Almagest. Thus Hipparchus' ideas became known to the modern world as the Ptolemaic system. Sic transit gloria mundi.

The Ptolemaic system was cumbersome, but amazingly accurate. But with thousands of stars to examine, astronomers still worried about the wanderings of those special five. On such nagging details are built the true glories of human thought. A Polish cleric, Nicholas Copernicus (to use his Latin name) suggested in 1543 that the ancient system of the heavens might be wrong. Perhaps the Earth itself is a planet, and all the planets revolve around the Sun.

Men died for espousing such views. Copernicus himself would not allow his ideas to be published until he was on his deathbed. But in 1609, when cantankerous Galileo Galilei first turned a telescope to the night sky, he saw that the planets were nothing at all like the fixed stars. The Moon was rough and mountainous. Venus showed phases as the Moon does, just as Copernicus predicted. Jupiter had four moons orbiting around it. There was something around Saturn's middle, too, but Galileo's telescopes were too poor to resolve what later was discovered to be a set of startlingly beautiful rings.

The telescope showed that the planets were worlds in space, more like our own Earth than the brilliant but unutterably distant stars.

Over the next three and a half centuries, astronomers spent most of their energies studying the five known planets and discovering new ones, some of which could not be seen with the unaided eye: Uranus, Neptune, Pluto, and the myriad small planetoids (often called *asteroids,* meaning "star-like") that orbit mainly between Mars and Jupiter. The Copernican system was not only verified, but thanks to the work of Galileo, Kepler and Isaac Newton, it led to a complete revolution in all of scientific thought. The concepts of motion, gravity, and the universality of physical laws all stemmed from Copernicus' humble idea of making a neater explanation for the migrations of the five naked-eye planets.

By the beginning of the twentieth century, astronomers had learned a good deal about the sizes, shapes, orbits, masses, densities, and rotation rates of the planets. But that was like trying to describe a distant cousin by listing his birth date, height, weight and clothing size. Not very satisfactory. There were many more unknowns about the planets than knowns. One of the major reasons for this frustration was that it was impossible to see any of the planets close-up; even in the best telescopes no planet could be seen as clearly as a naked-eye view of the Moon.

Astronomers turned their curiosities to the stars. For several reasons. First, they had more or less exhausted all the possible means of studying the planets, wrung them dry. Second—as the following chapters will show—many astronomers became embarrassed by the "popular" descriptions of life on other planets, descriptions that were at best enormously sanguine, at worst hoaxes, and at no time based on firm scientific evidence.

The popular press had seized on the pronouncements of astronomers such as Giovanni Schiaparelli and Percival Lowell, concerning sightings of "canals" on Mars, and trumpeted all over the world the news that Mars was inhabited by intelligent creatures who had built planet-spanning aqueducts. The result was a tremendous upwelling of interest among the general public, fanned by works of fiction such as H. G. Wells' "War of the Worlds," in which the super-intelligent but cold-blooded denizens of a dying Mars invade the Earth for *lebensraum*.

The general public might have been "turned on" by speculations about life on Mars and other worlds, but the astronomers for the most part were turned off. They found it impossible to conduct serious, sober research in a circus atmosphere. Every time an astronomer reported that he could find no evidence for the canals on Mars, he was drowned out by those who followed Lowell's optimistic reports.

Astronomers fled from the planetary studies, into stellar astronomy. More importantly, perhaps, new technical instruments such as the camera and the spectroscope gave astronomers powerful tools for investigating the stars. These tools were not very useful in studying the planets.

So a strange paradox took hold in astronomy. Telescopes became bigger and bigger, studies were pushed further and further into the limitless universe of stars and galaxies, and our neighboring planets were largely ignored. Certainly they were no longer in the forefront of astronomical research. Astronomers searched light-years beyond the fringes of the solar system, then millions of light-years and finally billions. Physicists bent their efforts to examining the subatomic processes that made the stars shine—and discovered nuclear energy in the bargain. Cosmologists evaded questions about the origin of the solar system, to construct theories about the origin of the universe.

Planetary astronomy withered.

But the excitement of seeking life on other worlds would not die. A new generation of writers, inspired by Wells, began a specialized field of publishing called science fiction. Young readers saw speculations about extraterrestrial life—often luridly exaggerated into "Bug Eyed Monsters"—splashed across the pages of popular magazines.

The real breakthrough came, though, neither from the astronomers nor the science fiction writers. It came from a quiet physicist named Robert H. Goddard, in 1926. In a snowy field near Worcester, Massachusetts, he launched the world's first liquid-fueled rocket. A new phase of history—and of astronomy—began at that moment.

By 1960, planetary astronomy was rekindled, thanks to the enthusiasm that had been kept alive in part by the writers and readers of science fiction, and to the enormous technical developments of new astronomical instruments. But the flame that really re-lit the torch of planetary astronomy came from the bellowing exhausts of rocket boosters. At last there was a new way to get more information about the planets—close-up!

By 1960 artificial satellites and space probes were garnering headlines all over the world. The famous Space Race between the United States and Soviet Russia was in full swing. Russia's *Lunik III* had photographed the far side of the Moon, the side that no human eyes had seen before. Astronomers began to realize that rockets could launch instrumented spacecraft toward the nearer planets, for close-up views of the alien worlds.

The first attempts to send instrumented probes to another planet, Venus, were made in 1961. They failed, and even the later successes were all but drowned out in the titanic glare of publicity given to manned space flights and the Apollo lunar landing program. But the rebirth of planetary astronomy went on, nevertheless.

In December 1962, *Mariner 2* flew past Venus. For the first time, we (meaning you and me, the whole human race) were able to get information about another world in space from close-up to that planet. Information that could never have been obtained from Earth, or even from a spacecraft in orbit around the Earth.

By 1971, our technical skills had reached such a level of finesse that we could place *Mariner 9* into orbit around Mars. And in the thirteen months from December 1973 through December 1974—the grandest period of exploration in all of human history—we explored the giant planet Jupiter, veiled Venus, and sun-scorched Mercury.

As this is being written, late in 1976, *Pioneer 11* has concluded its second flyby of Jupiter and is coasting outward toward a 1979 rendezvous with Saturn, the ringed planet, the most beautiful and enigmatic body in our solar system. And plans are being debated in NASA and the Congress to send new space probes even further, to the outermost limits of the solar system, to study the other worlds out there and see what lies beyond Pluto.

Since 1962, we have learned more about the Moon and four of the ancients' five wandering stars than we had been able to learn in the preceding century. Merely in the one year of 1974, more data on Jupiter and Mercury was obtained than all the information previously existing about all the planets.

This book is devoted to showing what these close-up probes of the new worlds have told us. For Mercury, Venus, Mars, Jupiter and our Moon are literally new worlds now—what we have found out about them since space exploration began has completely altered our view of their nature, and even forced scientists to re-think their ideas of the nature of our own planet Earth, and of the origin of the entire solar system.

Starting with the Earth-Moon double planet and progressing through Mercury, Venus, Mars and Jupiter, the first five chapters of this book show what space vehicles have told us about our own world and the other bodies they have visited. Each of these chapters discusses what was known about these worlds before space flight began, the often startling new information that the space probes have uncovered, and what the next steps in exploring these worlds will be. This book is not attempting merely to repeat what any reader can find in the newspapers or astronomy journals. The idea behind this book is to put the new discoveries into a context that will allow the reader to see what will happen in the future. This is needed, because the exploration of these new worlds has not reached its climax; it has barely begun.

Therefore, each study of the spacecraft-visited planets of the solar system in this book includes speculation on the future explorations and *human uses* of the new worlds. From dedicated solar observatories on Mercury to the discovery of alien life beneath the swirling clouds of Jupiter, the future expansion of human adventure and purpose throughout the span of the solar system is chronicled in this book, alongside the firmly factual accounts of what has already been learned about each planet.

Very few writers can deal both with detailed scientific fact and future-looking speculations based on these facts. Even fewer writers can handle both types of subject in an easily-readable, entertaining and understandable manner. We have a team of stars from the field of science fiction writing to handle this formidable task. As you will see from the biographical sketches given in each chapter, each of our writing stars is a scientist or engineer who has also written extensively for the science fiction audience. Each of them knows science thoroughly, can write entertainingly, and has the imagination to make fascinating speculations about the future of space exploration.

Two additional chapters deal with the distant parts of the solar system, beyond Pluto, that no human eye has yet seen, and with the latest scientific theories on the origin of the solar system. They too have been written by authors who have blended technical acumen with trained speculative imagination. All the material in every chapter of this book was written especially for this book, with the sole exception of "Styx and Stones: and Maybe Charon Too," which originally appeared in the November 1973 issue of *Analog Science Fiction-Science Fact* magazine.

In addition to our writing stars, this book is graced by the talents of Rick Sternbach, who painted the dust jacket illustration and the interior plates. He too combines the talents of a first-rate artist with the technical knowledge of an astronautical engineer.

From five wandering *planetos* that puzzled ancient thinkers, we are now sending our instruments and our intelligence throughout the solar system. Strangely enough, our own Earth is in a way the "newest" planet to be examined by astronomers. For it wasn't until spacecraft went into orbit that most human beings began to think of our own world as a planet—a tiny oasis of life and beauty in the vast and often hostile desert of space.

So welcome to planet Earth, and the other new worlds of astronomy's newest and grandest era.

Ben Bova
Trudy E. Bell
New York City
July, 1976

Closeup:
New Worlds

G. Harry Stine has been watching the sky as long as he can remember. His father, a Colorado Springs eye surgeon, bought him a 3-inch refracting telescope when he was 7, and he began scanning the clear Colorado night skies.

After receiving a degree in physics from Colorado College, he went to work at White Sands Proving Ground in New Mexico. During the daytime, he worked on rockets; at night he wrote about science and space. He learned planetary astronomy under the wing of Dr. Clyde W. Tombaugh, discoverer of the planet Pluto, and was president of the Las Cruces (New Mexico) Astronomical Society. When the Space Age opened with the beeping of Sputnik, he spoke up for a national space program... and promptly lost his job with a space rocket manufacturer.

Stine went on to found the National Association of Rocketry and Model Rocketry, to write nine books and hundreds of magazine articles, and to become a Fellow of the Explorers Club for chasing solar eclipses. He is a Fellow of the British Interplanetary Society, an Associate Fellow of the American Institute of Aeronautics and Astronautics, and a member of the New York Academy of Sciences. He has also been the director of a research lab for a major corporation, a product designer, science consultant to a major TV network, technological forecaster, and chairman of an international commission on rocketry.

Stine is currently the marketing manager of an instrumentation company and lives with his wife and three children in Phoenix, Arizona. He is a consultant to the National Air & Space Museum of the Smithsonian Institution and maintains one of the largest private collections of data and information on space craft and launch vehicles.

Sol III—The Twin Planet

G. Harry Stine

	TELLUS	**LUNA**
mean radius	6378 kilometers	1738 kilometers
mass†	1.000	0.012
mean density	5.52 grams/cubic centimeter	3.34 grams/cubic centimeter
surface gravity†	1.00	0.17
escape velocity	11.2 kilometers/second	2.4 kilometers/second
length of day*	23 hours 56 minutes	27 days 8 hours
length of year*	365.26	365.26
inclination of orbit to ecliptic	———	5°.1
inclination of equator to orbit	23°.4	6°.7
mean distance from Sun	1.000 AU = 1.49 x 10^8 kilometers	1.00 AU = 1.49 x 10^8 kilometers
eccentricity of orbit	0.017	0.055 (around Earth)
mean distance from earth		384,404 km

† Earth = 1
* Earth time

Swinging around its orbit third from Sol is perhaps the most unusual and most unique of all the solar worlds. It is properly called Tellus-Luna, although it goes by many other names locally. Since all other solar planets have been named after ancient Roman gods and goddesses, Tellus is the correct name for the major planet of the pair, since Tellus was the Roman goddess of the world, the protector and developer of sown seed. The other, smaller planet of the pair is Luna, Roman goddess of the months.

Tellus-Luna is a unique planetary system because it is the *only* twin planet in the Solar System. Although Luna is properly the satellite of Tellus because it is smaller, the

Tellus, the Blue Planet, our own Earth as photographed from the Apollo 17 spacecraft in December 1972. Beneath the swirling clouds are the southern Atlantic and Indian Oceans, separated by the continent of Africa. The point of Somalia and the Arabian peninsula can be clearly seen.

Luna, Earth's moon, photographed as the Apollo 17 *spacecraft approached it. Although there are four larger moons in the solar system, Luna is so large in comparison to its planet that the Tellus-Luna (Earth-Moon) system can be considered a double planet.*

size ratio between the two is more nearly equal than for any other solar planet and its satellite.

It is a unique pair, too, because of the great difference between the two worlds. Luna is a typical rocky planet, blasted with impact craters, devoid of any sensible atmosphere, and similar in appearance to both Mercury and Mars. Tellus is something of a freak among the solar planets; although it is also a small, rocky planet, it possesses a definite, thick atmosphere with an exceptionally high percentage of free oxygen, of all things. It is also the only planet in the Solar System, at this time, on which free liquid hydrogen monoxide—water—exists on its surface in exceedingly large quantities. Tellus is also unique because it is the only known solar planet which is known to support indigenous, self-aware life forms.

Tellus is the best-known planet because we live upon it. It is also the least-known planet because, until the Space Age, *we never saw it as a planet!*

This is reflected in the fact that the old names for our world—the Anglo-Saxon "Earth," the German "Erde," and the Latin "terra"—mean "dirt." But if the word for "dirt" makes you feel more comfortable when thinking about our home planet, use it.

Perhaps the best way to gain an appreciation of how the Space Age has changed our view of our own home world is to compare what we knew of Tellus B.S. (Before *Sputnik*) with what we have learned since.

For example, a 1927 astronomy book gives the following basic data about Tellus; it gives *only* this data, which leads to the obvious conclusion (the proof is left to the student as an exercise) that this is the only *important* data that one needs to know about the planet. Tellus is described as an "oblate spheroid" with a polar diameter of 7899 miles and an equatorial diameter of 7926 miles. (The metric system was not used to describe the dimensions.) It was reported to have a surface area of 197,000,000 square miles, 54,000,000 square miles of which were land not covered by water. The density of the planet is given as 5.6 times that of water. Its composition is described as "solid throughout and much stronger than steel." The astronomy book fails to mention earthquakes, faults, composition of the atmosphere, meteorology, geology, and a host of other subjects that today, A.S., are considered to be the "earth sciences." Ecology was a word to be found in unabridged dictionaries; demographics and other mod-

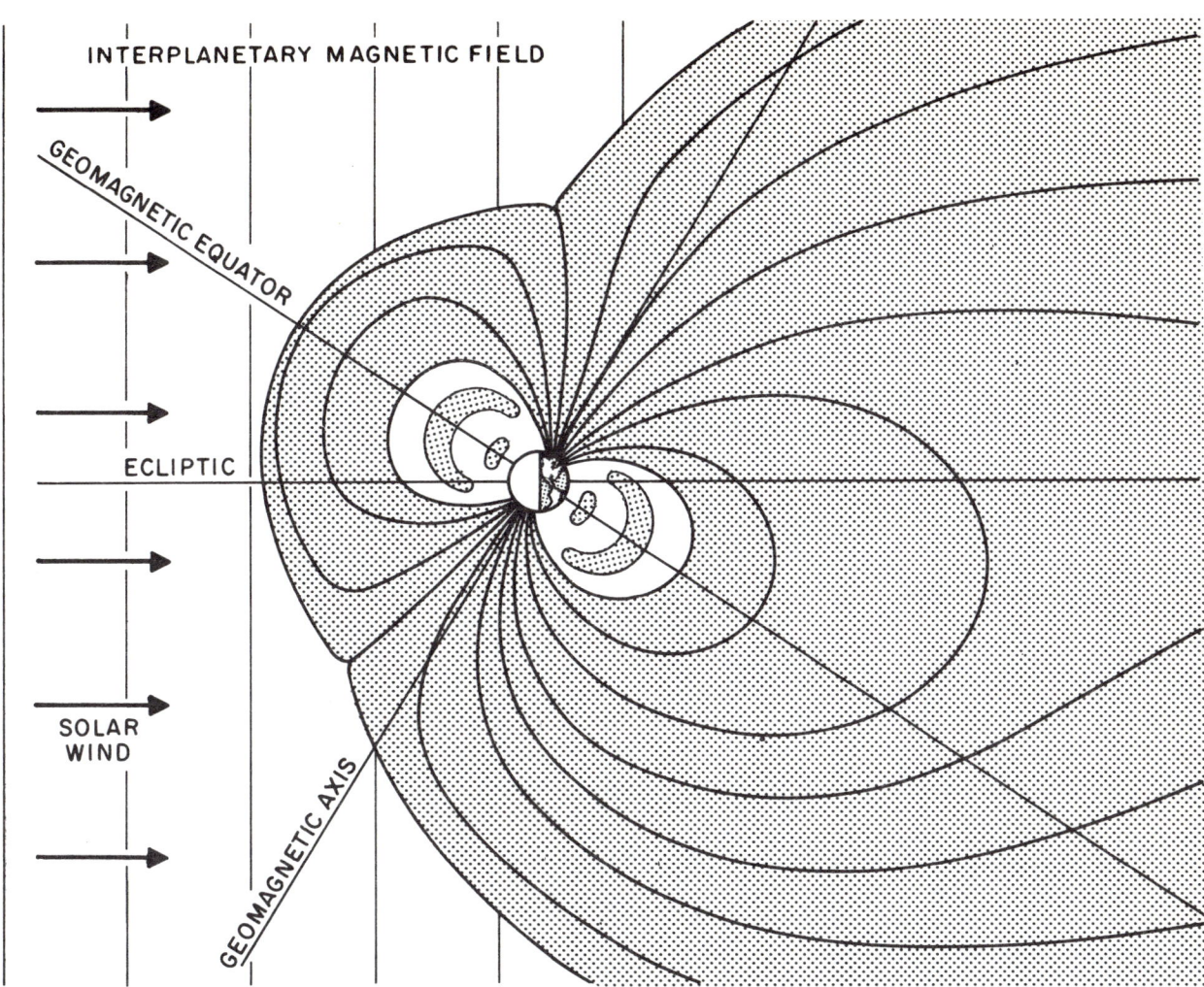

The Earth's magnetic field interacts with the solar wind and the interplanetary magnetic field to form an active magnetosphere, which is compacted on the sunward side, due to the pressure of the solar wind, and billows out well beyond the orbit of the Moon on the night side.

ern studies that relate the human race to the planet on which it lives either did not yet exist or were interesting little problem areas for graduate students to dabble in and about.

Meteorology was a local phenomenon. There was no global weather reporting or forecasting because there was no way to get weather data from remote land sites or from much of the ocean surface. Atmospheric electricity was a field yet in a descriptive stage although Nikola Tesla and others had looked into it in some detail. Air mass analysis existed more as a theory than anything else. The numerous interactions between the troposphere and the stratosphere were guessed at, but there was no theory whatsoever relating activities in the upper atmosphere to weather phenomena in the lower atmosphere.

It was known that the planet had a magnetic field, and it had been carefully measured and plotted because the difference between celestial north and magnetic north was of the greatest importance to aerial and marine navigation. But the B.S. concepts of the tellurian magnetic field seem amusingly simplistic to us today. Pre-*Sputnik* drawings of the planetary magnetic field show the planet "Earth" as a giant magnet with magnetic lines of force sweeping outward into space from both poles in great symmetrical loops. On the surface localized magnetic anomalies were known and plotted, but there was little knowledge of their cause.

Over both planetary poles, an upper atmospheric phenomenon called the aurora (Latin for dawn) was observed and photographed. It was not a true dawn, but an illumination of the upper atmosphere by unknown factors and causes. Auroral activity correlated to some extent with sunspot activity, but the reason for that was not known. The auroras were often accompanied by "magnetic storms" that disrupted and garbled long-range radio communications. These magnetic storms had a strong disturbing effect upon the "Kennelly-Heaviside Layer," a region about 80 miles above the planetary surface that reflected most of the low-frequency radio

Astronaut Edwin E. "Buzz" Aldrin, Jr., deploys the Passive Seismic Experiments Package (PSEP) on the lunar surface during the historic Apollo 11 mission in July 1969. PSEP recorded "moonquakes" that helped scientists to determine the inner structure of the Moon.

6

waves back to the ground, thus making most long-distance radio communication possible. This strange and inexplicable layer disappeared during daylight hours and re-appeared locally shortly after sunset. Listeners tuning their low-frequency radio receivers could detect strange sounds called "whistlers" whose activity increased near sunrise to provide a cacophonous "dawn chorus" of amplitude-modulated low-frequency radio signals.

Beyond the atmosphere, beginning 50 to 100 miles above the planetary surface, "space" began. Space was a pure, hard, cold, quiet vacuum. Nothing happened there. It was changeless. If you traveled far enough in the proper direction, you would eventually reach Luna.

The quiet vacuum of space was matched only by the quiet planet Tellus below. Compared to what we know today, "Earth" was a quiet planet. There was volcanism. There were earthquakes. There were tsunamis or "tidal" waves. There were landslides. And there was ample visible evidence of enormous forces that had, at various times in the past, convoluted and torn the surface of the planet.

Then *Sputnik*, *Explorer*, and *Vanguard* were injected into orbit to swing beeping around and around the planet. These early primitive spacecraft were soon followed by others of growing complexity, sophistication, sensitivity and selectivity. People followed. And for the first time we

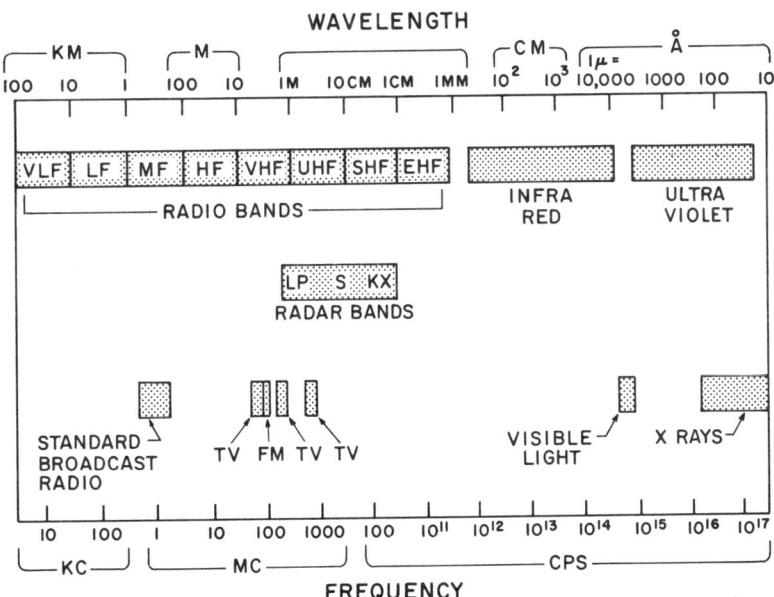

The electromagnetic spectrum.

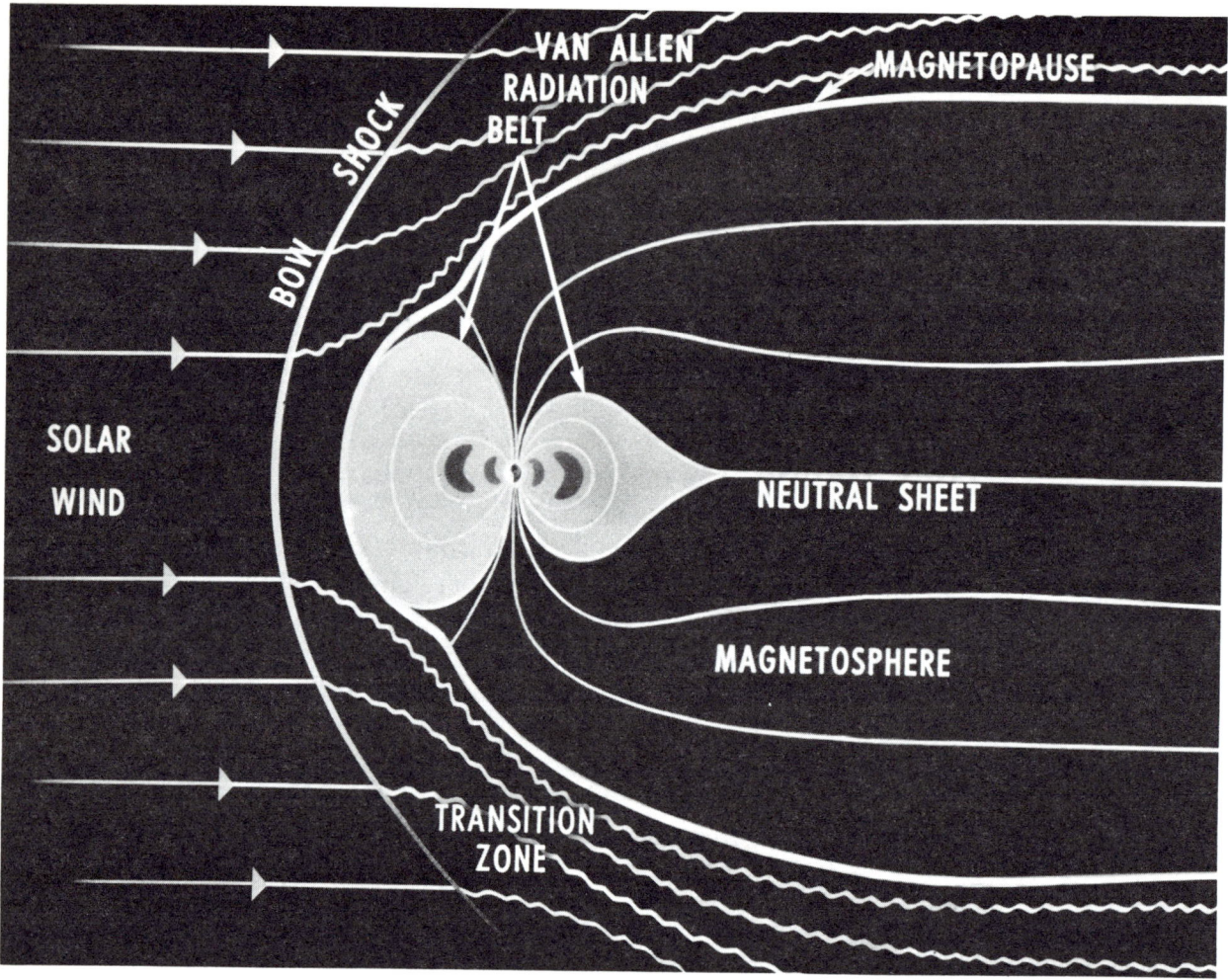

saw our home world through human eyes, cameras and sensors. What we have seen has given us a new view of our home world.

We now know that the atmosphere of Tellus no longer vanishes into the hard nothingness of space. Its pressure falls off nearly exponentially until it merges into the highly tenuous atmosphere of the sun, for we have learned that our planet orbits within the detectable atmosphere of a star. True, the pressure isn't very great; at an altitude of 500 kilometers, it has been measured at about 8.22×10^{-9} torr. A "torr" is 1/760th the atmospheric pressure at sea level on the planet. The composition of the atmosphere remains reasonably constant for the first 30 kilometers or so—21 percent oxygen, 70 percent nitrogen, and a conglomeration of carbon dioxide, water vapor, and the noble gases. Above 30 kilometers the incoming energy of

The geomagnetosphere, showing the shock wave, transition zone and magnetopause resulting from the "collision" of the solar wind against the Earth's magnetic field. Also shown are the Van Allen Belts, where high-energy electrons and protons are trapped by the field.

the sun changes things. It creates an ozone layer. Higher, it dissociates the molecules of atmospheric gases and, even higher, ionizes them (strips them of electrons). The ionization changes the density of electrons in the atmosphere and thus changes its electrical properties. The Kennelly-Heaviside Layer turned out to be more than a distinct layer in the atmosphere; it was a region that is now called the ionosphere. The change in its electrical properties is what causes electromagnetic radiation at radio wavelengths to be mirrored back to the surface below; as the energy input of the Sun changes daily, the electron density changes and so do the properties of the ionosphere.

The exact density of the upper atmosphere was initially measured by the first satellites, but *Vanguard 1*, launched 17 March, 1958, behaved queerly. Its orbit was not constant; it sped up and slowed down. The only thing that could make the satellite act that way was if the gravitational field of the Earth differed from one place in its orbit to another. That was due to the shape of the planet. *Vanguard 1* taught us that the planet Tellus is pear-shaped. In a diameter of about 12,800 kilometers, the antics of *Vanguard 1* revealed that there was a high-point at the North Pole and a bulging middle in the southern latitudes that averaged a 7.5 meter difference from a spheroid! That's one part in about 1.67 million.

Another early satellite, *Explorer 1*, discovered that orbital space was far from nothingness. It detected the Van Allen Radiation Belts, doughnut-shaped regions of high radiation around the planet. Other satellites refined these data, added more, provided new data. Slowly, a new view of Tellus in space was forged.

Now we know that the magnetic field of Tellus doesn't sweep outward into space in great loops like a huge bar magnet. It is distorted because our planet is a child of Sol. It is part of the Solar System. Immersed within the solar atmosphere, it is continually bombarded with a rain of energetic charged particles from the Sun. This solar wind distorts the planetary magnetic field, creating a "bow wave" or shock wave in front of Tellus and a wake behind, much like that of a boat or a supersonic airplane. The incoming solar particles, caught in the tellurian magnetic field, are trapped in a series of distorted doughnuts, the Van Allen Radiation Belts, around the planetary equator. When these charged particles collide with something in orbit—a space craft or an astronaut—they produce radiation as a result of their collision by a process called *bremsstrahlung*, the same phenomenon that creates X-rays for medical use on Tellus. The charged particles finally escape their magnetic trap by spiralling down the magnetic lines of force to the planetary surface in the vicinity of the tellurian poles. As they interact with the ionized molecules of the tenuous upper atmosphere, they produce a glow such as that in a neon bulb, forming the enigmatic auroras.

At the same time, the same charged particles from the sun, trapped in the magnetic field, can bounce back and forth from north to south while spiralling around magnetic lines of force, thus producing the strange "whistlers."

Our vantage point in space has also permitted us to study the heat balance of the planet, to learn its gross geology, and to discover that it is far from a quiescent world. It is a changing, dynamic, restless planet still alive and changing.

In 1912 Alfred Wegener put forth the concept of continental drift, the idea that all Tellurian continents were once one large land mass called "Pangaea." It was an attractive hypothesis because anyone with a map or globe of the world can see at once that the east coast of South America seems to be a perfect fit into the west coast of Africa. But it was difficult, if not impossible, to get a fit anywhere else. Today, Wegener is recognized as the pioneer in the new "plate tectonic" theory of the movement and formation of the surface of Tellus. But to get from Wegener's 1912 theory of continental drift to plate tectonic theory required undersea exploration, better measurement of the planet, comparisons of rock samples and fossils from different mountain ranges separated by oceans, and the development of the new field of paleomagnetism.

Today's plate tectonic theory developed not only from the increase in the activity of the new "earth sciences" brought to life by a new view from space, but also from our space studies of Luna and Mars. In studying other worlds, we began to better understand our own.

Part of the Salton Sea, in California, photographed by the Earth Resources Technology Satellite (ERTS) in 1974. Sensors aboard ERTS and similar satellites can provide information valuable to farmers, geologists, city planners, ecologists, and pollution control agencies.

Plate tectonic theory hypothesizes that the surface of a planet is made up of crustal plates floating on a semi-liquid magma. These plates range from 8 kilometers thick beneath the oceans to as much as 50 kilometers thick at the core of a continent under mountain ranges. There are eight major continental plates and several minor ones. They wander over the surface of Tellus propelled very slowly by convection currents in the liquid magma beneath them.

It is those same convection currents in the molten core of Tellus that probably create the unique planetary magnetic field of Tellus. None of the other three inner planets of Sol, nor Luna itself, has as strong a magnetic field as Tellus.

Where two tectonic plates are spreading apart, a rift valley appears with a vein of volcanism; the valley continues to spread the plates apart. It's happening right now in the middle of the Atlantic Ocean. Spacegoing measurements from satellites have revealed that America and Europe are growing further apart by a matter of centimeters every year. Where two crustal plates collide, they may buckle upward to form mountain ranges, or one may slip under the other to form a trench.

Plate tectonic theory has provided us with a new explanation of most earthquakes and may, in time, form the theoretical basis for earthquake prediction.

Although this is part of our new view of Tellus as a dynamic, changing planet, plate changes take place over very long periods of time. For example, it will take nearly 50,000,000 years for Los Angeles and the part of California west of the San Andreas fault to move northward and disappear into the Aleutian Trench... or for the "New Himalayas" to form from the continuing collision of the Arabian plate with the Asian plate.

We have a new view, too, of the insides of our planet. The part of Tellus that we can see, experience, or visit ranges from the upper fringes of the atmosphere at 300 kilometers or so above sea level down to the deepest ocean trenches about 11,000 meters below the surface. There is much more that is yet inaccessible. Below the crust of Tellus is the liquid mantle some 2,900 kilometers thick, consisting of rock heated to temperatures as high as 4,000° C. The final 3,500 kilometers to the center of the planet is the core, liquid at its outermost part and solid at the center. The core is primarily iron, and motions of eddy currents within it creates the unique planetary magnetic field of Tellus.

It turns out that Luna is quite different from Tellus in many ways. Our Space Age voyages to Luna have uncovered evidence that the two planetary bodies probably have different origins. Before any instruments or astronauts had reached Luna, there were three major, basic theories of its origin. They may be called the "daughter," "sister," and "spouse" theories.

The "daughter" theory maintains that Luna was born from Tellus, that the moon was ripped from the primordial proto-planet during the early eons of the formation of the Solar System.

The "sister" theory claims that Tellus and Luna were both formed at the same time together from the same cloud of primordial gas, that they have always orbited each other: that they are sisters, a true twin-planet system.

Lastly, the "spouse" theory argues that Luna was a separate planet all its own that was somehow captured by Tellus and wed to the big, blue planet forever by gravity.

In spite of the fact that the Apollo program succeeded in landing 12 men—only one of whom was a geologist—at six different locations on Luna and returned some 382 kilograms of lunar soil to Tellus for study, there is still no clear-cut answer as to which is the correct theory of the formation of the Tellus-Luna pair. As one scientist remarked, it would be easier to explain the absence of a tellurian satellite than to explain the presence of Luna.

The daughter theory suffers from the fact that the chemistry of Luna seems to be different from that of Tellus. Samples of lunar soil and rocks show that the satellite is rich in the so-called refractory elements, those with high melting points such as calcium and titanium. Granite of the kind that is common on Tellus—a metamorphic mixture of feldspar, mica, and quartz—seems to be rare on Luna; there are no granite outcrops on Luna that are comparable to those on the tellurian continents. Although metallic iron is common in the Apollo lunar samples, the iron may well have come from the explosions from the impacts of meteorites during the early eons of lunar history.

From the samples thus far returned to Tellus for study, selenologists have concluded that there are only four major types of lunar materials. First, there are the basalts from the lunar plains that are similar to the volcanic rocks

of Tellus; these could very well be from the lunar mantle. Secondly, the lunar highland areas seem to be covered with a type of rock called "anorthosite" that is rich in feldspar, a silicate of aluminum and calcium; according to *Apollo-17* geologist-astronaut Jack Schmitt, the anorthosites are both uncommon and extremely old on Tellus. The most unusual lunar rock has been given the apellation KREEP—K standing for potassium, REE for rare earth elements, and P for phosphorus—found primarily in certain lunar lowland areas; it has a high percentage of radioactive elements. Finally, in the Descartes region of Luna is found a basaltic-type rock with very high aluminum content, totally unlike any tellurian rock.

Granted that although a limited lunar sample obtained from only six landing sites isn't really enough to offer an open-or-shut case for lunar minerology and chemistry, it appears difficult to account for the daughter theory or even for the sister theory. If both planets had the same origin in the same proto-planet or gas cloud, one would expect them to have reasonably similar chemistries. One modification of the sister theory, which might be called the "niece" theory, suggests that Luna accreted from a Saturn-like ring around Tellus that was of slightly different composition than the material that formed the primary planet.

The big problem with the spouse theory is the difficulty in coming up with suitable orbital dynamics that would allow the larger planet, Tellus, to capture and hold Luna. At the present time there is no suitable explanation for how it could have happened. On the other hand, none of the lunar studies to date have produced any evidence to rule out any of the three theories! The true story is still an enigma.

Scientist-Astronaut Harrison H. Schmitt working beside a boulder on the Moon during the Apollo 17 *mission in 1972. The lunar rover is in the foreground.*

The astronauts picked up enough rock and soil samples, however, to permit an accurate estimate to be made of the age of Luna. The samples indicate that most of the lunar surface is 3.16 to 4.3 billion years old. Compare this to the fact that about 99 percent of the surface of Tellus seems to be less than 3.1 billion years old. Of course, Tellus has been subjected to an unusual level of erosion for those billions of years—wind, thermal, and water. The motion of the tectonic plates has probably erased most of the original tellurian surface. On Luna the surface has lain bare to space for all those billions of years, acted upon only by the erosion of the temperature cycle of lunar night and day, the action of the solar wind, and impacts of meteorites.

But each of these three types of lunar erosion was grossly misjudged by lunar experts before the Space Age. For example, conceptual paintings of the lunar surface before the Apollo landings showed ranges of steep, sharply-spired mountains. We now know from on-the-spot photographs that the highest and most rugged lunar mountains are gently-sloped, softly-rounded hills. At one time, they may have been rugged. But some three billion years of erosion from the day-night temperature cycle—ranging from 110° C. at lunar noon to -179° C. just before lunar dawn—plus billions of years of countless impacts from micrometeorites have reduced those once-majestic mountains to ant hills.

The lunar surface also shows the signs of more than three billion years of meteoric impacts. Water erosion on Tellus has erased nearly all the evidence of meteorite impacts except for geologic fossil craters. On Luna, the bombardment began soon after its formation and went on for perhaps a billion years thereafter. The most massive impact on Luna was almost four billion years ago when a large planetoid that may have been as much as 100 kilometers in diameter slammed into it. It formed a basin more than 1,000 kilometers in diameter that today we call Mare Imbrium; it's the right eye of the "Man In The Moon" that you can see with the naked eye. Because this planetoid impact punched right through the very hot crust and into the lunar mantle, the Imbrium basin soon filled with molten lava to become as we see it today.

Luna is still undergoing meteoritic bombardment. On 17 July, 1972, a meteorite almost two kilometers in diameter struck the side of Luna away from Tellus. The

Astronaut Schmitt studying a huge split boulder at the Taurus-Littrow area explored during the Apollo 17 *mission.*

Mount Hadley, in the lunar Apennine range, rises 14,765 feet (4429.5 meters) above the plain. Photographed during the Apollo 15 mission in 1971.

impact was picked up and recorded by the ALSEP stations—*Apollo Lunar Scientific Experiment Packages*—left on the lunar surface by the Apollo astronauts. One of the instruments in each ALSEP is a seismometer to measure lunar tremors—moonquakes, if you will. Hundreds of such have been detected. Other instruments in the ALSEPs measure surface temperature and the tenuous lunar atmosphere.

The six ALSEP units on Luna have shown that this world is not inert as once thought; it is still an active, dynamic, changing celestial body. ALSEP was originally designed to last for one tellurian year; scientists have been elated over the fact that, after more than five years on the lunar surface, five of the original six ALSEPs are still working.

ALSEP thermal measurements indicate that Luna is still active thermally: heat is flowing out from its interior. The heat-flow data coupled with the seismometry data have given selenologists a new look at the make-up of the lunar interior.

The lunar crust seems to be about four times thicker than that of Tellus; there is no known reason for this except for the possibility that Luna cooled more rapidly because it has less mass than Tellus. No evidence of recent volcanism has been revealed by ALSEP measurements.

The lunar mantle appears to extend inwards about 1,000 kilometers with a temperature building toward 1,700° C. The mantle merges into a partially-molten lunar core extending the final 770 kilometers to the center. The lunar core does not seem to be nickel-iron as that of Tellus is. The core is probably heated by the decay of radioactive elements within it.

The lack of a nickel-iron core partially explains another finding of the Space Age: the almost total absence of a lunar magnetic field. Without a molten magnetic core that is in motion, there is little reason to suspect a magnetic field or even plate tectonic activity as there is on Tellus. And, indeed, there seems to be little plate tectonic activity, because the lunar surface is now much the same as it has been for four billion years.

Another piece of evidence that indicates that Luna does not possess much—if any—nickel-iron core comes from the fact that the density of Luna is about 60 percent the density of Tellus. Since the surface and crustal rocks on Luna have about the same density as those of Tellus, and assuming that the lunar basalts have about the same density as the mantle rocks, Luna's low overall density would indeed tend to confirm a lack of a large nickel-iron core.

How was the lunar density measured? It was determined primarily by satellites that we put in orbit around Luna such as the *Lunar Orbiter*. Precise tracking of the orbits of these lunar satellites permitted an accurate determination of the lunar gravitational field, which in turn is a function of the lunar mass. Since the diameter of Luna is known, its density—mass per unit volume—could be calculated.

The lunar orbiters revealed other things as well. Just as the tellurian satellites led to the discovery of a new shape for Tellus, the orbits of the lunar satellites showed that Luna is lop-sided. Its center of mass is far off from its center of volume. The center of mass of Luna is displaced toward Tellus by almost three kilometers. When the larger, more sophisticated Apollo spacecraft orbited Luna, their sensitive instruments confirmed this fact.

The lop-sidedness of Luna is due to the gravitational pull of Tellus on it, locking it into position so that its one face is always turned toward Tellus. There is not, however, any theory to explain some of the other ways in which the satellite is non-symmetrical:

The "far side" of Luna—the side that is always pointed away from Tellus—is almost devoid of the large, lava-filled *maria* of the "near side." The far side is covered with impact craters except for a few small, lava-filled basins. A hypothesis has been put forth to explain this: the lunar crust may be thicker on the far side by as much as 50 kilometers. In answer to the question, "Why is the lunar crust thicker on the far side?" the answer is a non-committal shrug at this time. We don't know why. Nor do we know why lava in large quantities failed to fill some of the basins on the far side as it did on the near side.

Orbiting lunar satellites also discovered that most of the surface radioactivity on Luna seems to be concentrated in a few distinct areas on the near side with very little on the far side. The areas seem to be concentrated in locations near major impact craters and maria—Copernicus, Kepler, Aristarchus, Mare Nubium and Mare Serenitatis, for example. Again, no one knows why this is so. And no one knows whether or not these areas might contain major deposits of radioactive ores. After six manned landings, we realize that we know pitifully little about this lunar world that accompanies our planetary home.

Lastly, orbiting lunar satellites did not maintain a constant orbital velocity; *Lunar Orbiter 5* speeded up temporarily when passing over certain portions of the lunar surface. This increase in speed wasn't much—about one part in ten thousand—but it was measurable. And it led to the discovery of the lunar "mascons," the word being shorthand for "mass concentrations." The mascons are more frequent on the side of Luna facing Tellus and their presence is apparently correlated with the large impact basins. Again, we don't know what the mascons are or what caused them. The best guess to date suggests that the mascons are caused by an up-welling of the lunar mantle or perhaps of materials from the core to replace the crustal material blasted away by the impact of the large bodies forming the basins early in the lifetime of Luna. This material may be of higher density than the original crustal material.

Possibly a synthesis of the data about the lopsidedness, the concentrations of radioactivity, and the mascons might someday in the future provide the answers.

Although the majority of the lunar craters now seem to be of impact origin instead of volcanic, thus settling a long-standing pre-Apollo controversy among lunar astronomers, we are still hard-put to explain other common features on the lunar surface. For example, the sinuous rills still remain a major puzzle, even though *Apollo 15* landed almost astride one. These rills are wandering valleys on the lunar surface. Yet, because there is no water on Luna, the rills are not caused by the erosion of running water. Are they subsurface lava tubes or vents? Are they the result of subsidence where "the roof fell in" in some manner? We don't really know yet.

We have seen much on Luna that correlates with what we see on Tellus. Yet there is much on Luna that we have never seen before and that we cannot explain in tellurian terms. We have come to realization that we have a distorted notion of the laws of physics, Sir Isaac Newton to the contrary. We have all been born and grown up in a one-*g* gravity field. We have been able until recently to study only one planet, and we have had to study it while surrounded by a one-*g* field and our planetary atmosphere. The geological processes that we have observed, deduced, hypothesized, and studied hold true here...on a small planet surrounded by an oxidizing atmosphere at a pressure of 14.7 pounds per square inch (1.036 kilograms per square centimeter) under conditions that permit the existence of free liquid water. This has distorted our view of planetology. Geological processes do not have a one-to-one correlation with selenological processes.

For example, volcanism on Luna produces an end result that appears quite different from tellurian volcanism. On Tellus, we see volcanism actively at work; on Luna, all volcanism seems to be a matter of history. On Tellus, we see volcanoes and large calderas; on Luna, we see what may have been very large calderas at one time in the far past, but we have yet to see an active volcano.

There have been transient phenomena observed on the lunar surface through tellurian telescopes. The craters Aristarchus and Grimaldi have been sites of chimerical spots, sometimes glowing with faint color. These have been reported since the 16th century. In 1963, professional observers at Lowell Observatory in Flagstaff, Arizona observed three reddish patches near the crater Aristarchus that persisted for over an hour.

These glowing patches may be due to what latent volcanism remains on the lunar surface. Orbiting lunar satellites have detected traces of radon gas in the vicinity of Aristarchus. This radioactive gas may be seeping out of the lunar interior.

There have been no permanent, visible changes in the physical features of the lunar surface since astronomers began to map Luna centuries ago. The only possibility of a slight change is in the crater Linne, which seems to have changed from a mountain to a bright crater over a period of a century or so... if we can really believe some of the earlier lunar observations made with optical instruments that may or may not have been up to modern quality.

Thus, we have learned that we cannot adequately explain or answer some of our questions about Luna from the data we now have on hand. And, although we have discovered that we can more readily study, observe, and learn about our own planet, Tellus, by observing it from orbital space, we find that there is much that we do not know about our own home planet, too.

Should we answer these nagging questions about Tellus and Luna? Or should we simply note that it has been an interesting exercise for a few scientists and then return to our ground-bound pursuits?

Geology becomes selenology as Scientist-Astronaut Harrison H. Schmitt removes a sample of lunar rock for return to Earth, during the Apollo 17 mission.

Soil samples were also collected during the Apollo missions, and returned for study by scientists on Earth. Here Dr. Schmitt scoops up some of the sandy lunar soil in an implement specially designed for the task.

Our space view of Tellus has reinforced our growing knowledge that our planet is finite in size and resources. We have also discovered that our planet has a system of living things called an ecology. We know that the numbers of the human race are growing steadily and will continue to do so, barring any great leap forward in the area of planet-wide birth control and longevity. We know that we have used up much of our planetary natural energy sources and that we will have to start recycling some of our planetary raw materials. In short, we have been hit between the collective eyes with the fact that we are indeed riding along together on a big space ship. Thus far in our turbulent and somewhat rapacious history, we've been able to get along with rape and murder, so to speak. But it is dawning upon some of the human race that we've got to do a better job of managing our home planet.

And we know so little about it.

We don't really know the effect of releasing fluorinated hydrocarbons from aerosol cans into the atmosphere. We don't know the long-term effects of jet airliners flying in the stratosphere. We don't know what will happen if we do or do not continue doing many of the things we now do. And it is exceedingly risky to experiment with an entire world... particularly the one on which we live.

In this regard, learning more about our home planet by observing it from space with instruments that are improving day by day in sensitivity and selectivity cannot help but give us better information upon which to base decisions of technical, political, and social importance. The big danger is not that we will make the wrong decisions on the basis of such data, but that we may make disastrous decisions based on ignorance of the true facts of our planet.

But what of Luna? Are we justified in continuing to ask for answers to our questions about the other half of our twin planet system? What difference does it make whether Luna came from Tellus, was formed along with Tellus, or was captured by Tellus billions of years ago?

These questions will probably be answered only by long-term, on-the-spot human exploration of Luna. After all, how much would we know of Tellus, which has roughly four times the dry land area of Luna, if we had landed only 48 men, four of whom were geologists, at 24 points on the planet? It takes years of crawling across the

Our living world, as seen by a meteorological satellite. Hurricane Carmen is moving across the U.S. Gulf Coast (upper left, arrow) while South America, the Atlantic Ocean, the west coast of Africa and the Iberian Peninsula stand relatively free of clouds.

landscape on foot and in wheeled vehicles to complete a thorough geological survey of a single western state.

We cannot justify expending the time, effort and capital resources of returning to Luna and establishing long-term exploration bases simply to answer our questions of lunar origin for the sake of knowledge alone. However, our newly-found space view of Tellus has shown us that we live on a planet of limited size and resources; by means of space flight, we will begin to manage those tellurian resources better and begin to look elsewhere in the Solar System for sources of energy and raw materials to replace tellurian sources. The long-term exploration of Luna will probably be carried out in a search for lunar raw materials—radioactive elements for nuclear power sources, titanium, aluminum, and other basic metals of our planet-wide civilization. The questions of the origin of Luna will be answered as a "fall out" or windfall of this very pragmatic and useful exploration of Luna.

When Thomas Jefferson concluded the arrangements for the Louisiana Purchase from France in 1803, he opined that perhaps these newly-acquired lands might be fully settled by the year 2600 A.D. He sent Lewis and Clark to explore these lands in an early-day equivalent of today's space program in terms of purchasing power of the dollar and percentage of the gross national product consumed. Lewis and Clark walked right over the fantastic mineral wealth of the American West and never even suspected it was there. A half-century later, a Senator from Indiana felt that it would do no harm to provide the fledgling Western railroads with federal land grants because, "These lands can be classed as refuse," he stated.

On 25 December 1968, *Apollo-8* astronaut Frank Borman looked down on Luna from orbit and remarked, "It's not a very good place to live or work." Standing on that same lunar surface on 20 July 1969, Edwin Aldrin remarked, "Magnificent desolation."

Perhaps Luna is as desolate and "useless" as the endless prairies of the Great Plains must have seemed to Lewis and Clark. We have yet to find out. We have yet to really explore Luna. And we probably will because *we need it*.

Are we justified in continuing to ask for answers to the question of lunar origins? If these answers in turn provide us with better information on how to manage our home planet, or if they can provide us with additional resources, this quest will not only be justified but will also yield extremely valuable information... valuable to the tune of billions of dollars. And also valuable in terms of our outlook on the universe.

When animals are confined in a closed cage with limited resources and overcrowding, they go insane and begin to kill one another. Human beings are the only animals who do this deliberately to themselves in large cities and overpopulated regions. A foreseeable future based on a closed planetary cage with dwindling resources and increasing population naturally leads to a "doom and gloom" outlook. An open cage, an expanding frontier, and new worlds to conquer breed a healthy outlook on the future.

Increasing information about the twin planet system and the growing exploitation of that information and of the system is a definite and logical extension of what has been going on for most of this century. It does not take very much imagination to foresee that if we can keep people alive, happy, healthy and working in Skylab for 86 days as well as on the surface of Luna for hours at a time, we can keep people alive, happy, healthy and working in space and on Luna for whatever length of time necessary.

The colonization of the twin planet system from the larger body of the two has been the subject of a great deal of serious scientific and technical study for several decades. There are dozens of serious technical proposals on setting up a permanent lunar base. There are recent concepts of deep-space colonies located at the Lagrangian points of the Tellus-Luna system.

The Lagrangian points are a special solution for the nearly-insoluble problem of three-body celestial mechanics. It is a nearly-insoluble problem because we do not know how to compute the orbits of three bodies in space each of which gravitationally affects the other. With high-speed digital computers, we have been able to approximate the solution by a series of successive iterations. One of the special solutions is the arrangement of the three bodies in an equilateral triangle with two smaller bodies orbiting the largest one and the two smaller bodies 60 degrees apart from each other in the same orbit. We know it to be a valid solution because the Trojan asteroids lead and follow the planet Jupiter in the 60-degree location.

What has been missing from these concepts is a basic reason for doing it at all. As stated previously, we can't logically justify it on the basis of obtaining scientific data. We can only justify it on the basis of exploitation of the environment, the production of capital wealth, the relocation of basic, heavy-consumption, heavy-polluting industrial operations into the space environment to utilize new raw material sources in the twin planet system and the Solar System itself. I have done considerable work on the further development of this "third industrial revolution" concept. It is based upon the proven fact that when somebody can make a buck doing something, it will get done.

Thus, our new view of the twin planet system of Tellus and Luna not only includes a great deal of new information about these two worlds, but also some new outlooks on what they mean to our future. To say that information about our world and the way we look at it has no bearing upon our way of life belies all historic evidence. When our ancestors were trapped in Europe on a flat earth, they looked at life a great deal differently than we do today. What will our progeny have to think about us, trapped on a finite Earth in a hostile and useless Solar System? Or can we anticipate them a little bit?

We are a long way from complete understanding of our twin planet system. We know so little about its ecology and how we might change it or save it. Each answer we obtain seems to create more questions. Each new question seems to be more critical to our lives and our future. Each new space ship that leaves Tellus gives us a few more pieces in the jigsaw puzzle of where we came from.

And until we know where we came from, we certainly can't get a very good picture of where we are going.

There is perhaps one word that can sum-up everything we are learning about our new view of Tellus-Luna:

Perspective.

In both time and space.

Of our world and of ourselves.

JERRY POURNELLE is science editor of *Galaxy* magazine and a syndicated science columnist with the National Catholic Press publications.

He was an aviation psychologist and the Chief of the Experimental Stress Studies Project in the early days of space travel, and his ground simulation work in connection with *Project Mercury* earned commendations from both the *Mercury* and *Gemini* project leaders.

He was also manager of special projects studies for Aerospace Corporation, and was a team leader on several *Apollo* design studies. He left the aerospace industry to become the founder and first president of Pepperdine Research Institute, where he conducted studies under contracts to Headquarters, USAF, the Department of Transportation, NASA, and the Department of Justice. Since 1969 he has been a free lance writer and consultant, and briefly held the post of Executive Assistant to the Mayor of the City of Los Angeles and city Director of Research.

Dr. Pournelle lives in Studio City, Calif. with his wife and four sons. He is a past President of Science Fiction Writers of America, and a member of Aviation and Space Writers Association, the Institute for Strategic Studies, and the American Institute of Aeronautical and Astronautical Engineers. He was the first winner of the John W. Campbell award for the best new S-F writer of 1973, and is author of several science fiction and non fiction books.

The Old and New Mars

Jerry Pournelle

	MARS
mean radius	3394 kilometers
mass†	0.108
mean density	3.95 grams/cubic centimeter
surface gravity†	0.38
escape velocity	5.1 kilometers/second
length of day*	24 hours 37 minutes
length of year*	686.98 days
inclination of orbit to ecliptic	1°.8
inclination of equator to orbit	24°.0
mean distance from Sun	1.52 AU = 2.23 x 10^8 kilometers
eccentricity of orbit	0.093

satellites	diameter (km)	distance from planet (millions of kilometers)	orbital period (days)*	date of discovery
Phobos	18x22	0.009	0.31	1877
Deimos	12x13	0.023	1.26	1877

† Earth = 1
* Earth time

Since men first looked through telescopes Mars has been the most interesting of our neighbor planets. Our views of the fourth planet have undergone minor as well as radical changes, but there have been three major theories about the conditions on Mars. The first two were dead wrong. The last, our present theory, is incomplete.

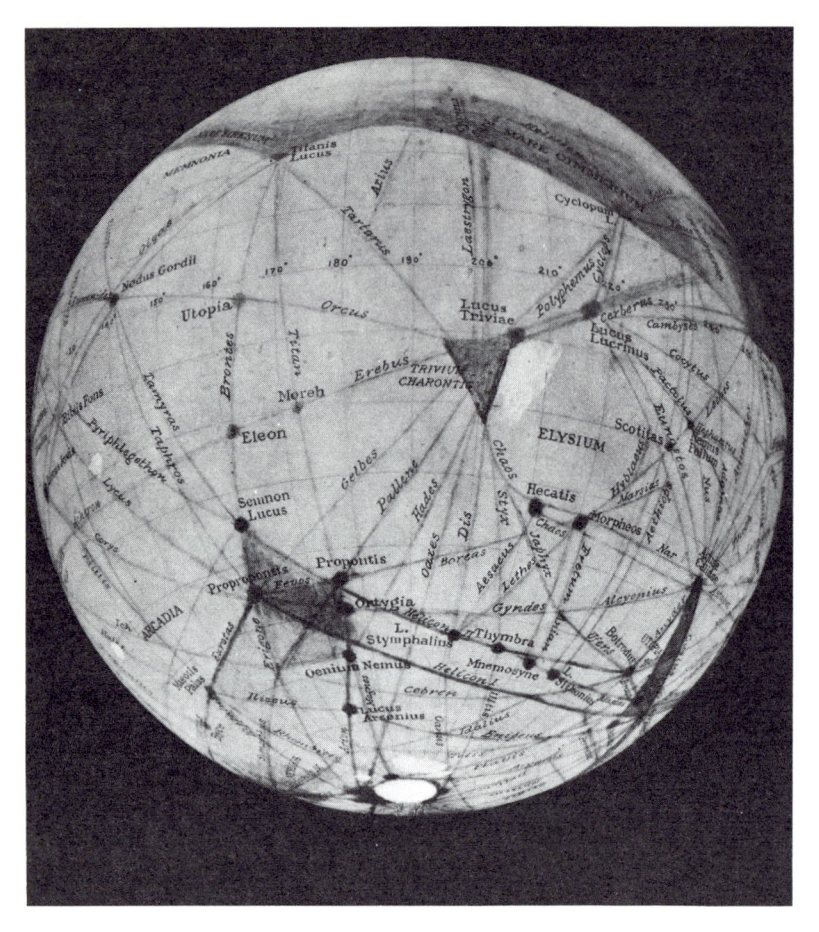

Percival Lowell's globe of Mars. Almost none of the marks and canals on this globe is of use in understanding the real Mars. Lowell was a dedicated observer, but he was limited by the instruments available to him. (Lowell Observatory)

We first thought Mars was like the Earth, and possibly held intelligent beings. Certainly, we thought, it had some form of life. Then, when the first Mariner spacecraft sent back photographs of cold, dead craters, we thought Mars was like the Moon. Now we know that Mars is a unique planet like no other, a dynamic world from which we can learn much about the Earth.

For all that, the Old Mars influences us yet. It was one of our most cherished astronomical theories, and played a crucial part in the development of our space program. The Viking spacecraft probes grew out of the popularity of the Old Mars, and their very orbits have been determined by expectations more emotional than scientific. If we look at what has been known about Mars all along—as opposed to what we believed—we should not be surprised that the earliest spacecraft found no evidence for life, much less signs of intelligent beings. Yet many were not only surprised but shocked. The Old Mars never existed, but it was a planet that many believed in deeply and passionately.

It isn't hard to see how this came about. Mars seen from Earth looks much the way we might imagine Earth would from a long way off. We have known since the 17th Century that the Martian day is very nearly the same length as our own. The inclination of its equator to the plane of its orbit is very nearly the same as Earth's. It has seasons during which surface features change. Even as seen through a small telescope the polar icecaps stand out brilliantly against the reddish surface, and large areas of dark and light suggest seas and continents. It is small wonder that Mars was thought by early observers to be Earth's twin brother.

No other planet seen from the Earth reveals surface details like those of Mars. Mercury shows only irregular smudges, Venus is an undifferentiated cloudy ball, and Jupiter and Saturn show changing bands and shifting spots. Only Mars can be mapped from the Earth. Furthermore, a great deal of the detail can be discerned with small telescopes. Larger telescopes, curiously enough, show little more: when you step up the magnification you also enlarge the infuriating shimmer caused by turbulence in the Earth's atmosphere, and you quickly find that the diameter of the telescope is not so important as the steadiness of the air. Even with the best seeing, however, the air has enough residual unsteadiness that details stay maddeningly beyond the limits of visibility. Just as you concentrate on some fascinating structure it vanishes in a blur. For a brief rare flash you perceive a wealth of incredibly fine tracery...and then it melts away and frantically you try to capture it all on paper before it fades from memory. Afterwards you may sit and stare for hours at what you've drawn. What does it all mean? Is there a pattern?

THE OLD MARS

Our picture of the Old Mars probably grew out of one of our first theories of the origin of the solar system, the nebular gas hypothesis of Immanual Kant and Pierre Simon LaPlace. They thought the planets formed from swirling clouds of gas, with the outer planets condensing before the inner ones. Thus Mars, according to the theory, was like the Earth but older, while Venus showed us what the Earth had been like when it was younger. It was an appealing image.

Then during the favorable opposition of Mars in 1877, Giovanni Schiaparelli, Director of the Brera Observatory in Milan, had a moment of superb seeing. He was using a fine instrument in a good location, and Mars was as close to Earth as the planet ever gets. Schiaparelli's observation at that moment started a controversy that lasted until the Mariner spacecraft put an end to it once and for all.

In that moment of excellent seeing the lighter regions of Mars seemed to be covered with a complex network of fine straight dark lines. They appeared to run for hundreds, even thousands, of kilometers, linking the dark "seas" and never starting or stopping in a light area of "land". Where the lines intersected there was a dark spot that might have been a lake. Schiaparelli named these lines *canali*, Italian for channels. He suspected that they might be the work of intelligence, but didn't insist on that.

No one paid him the slightest attention even when during the opposition of 1879 he saw them again and announced that they were permanent features. Other astronomers looked for them and failed to find them. Then, in 1888, French observers saw the *canali*. In rapid succession others began seeing them as well; they received good press and the Old Mars was born.

Mars might still have remained unimportant to astronomers had not Percival Lowell, brother of the poet

A modern map of Mars, prepared at the University of Texas for the Jet Propulsion Laboratory's planning of the Mariner 9 *mission.*

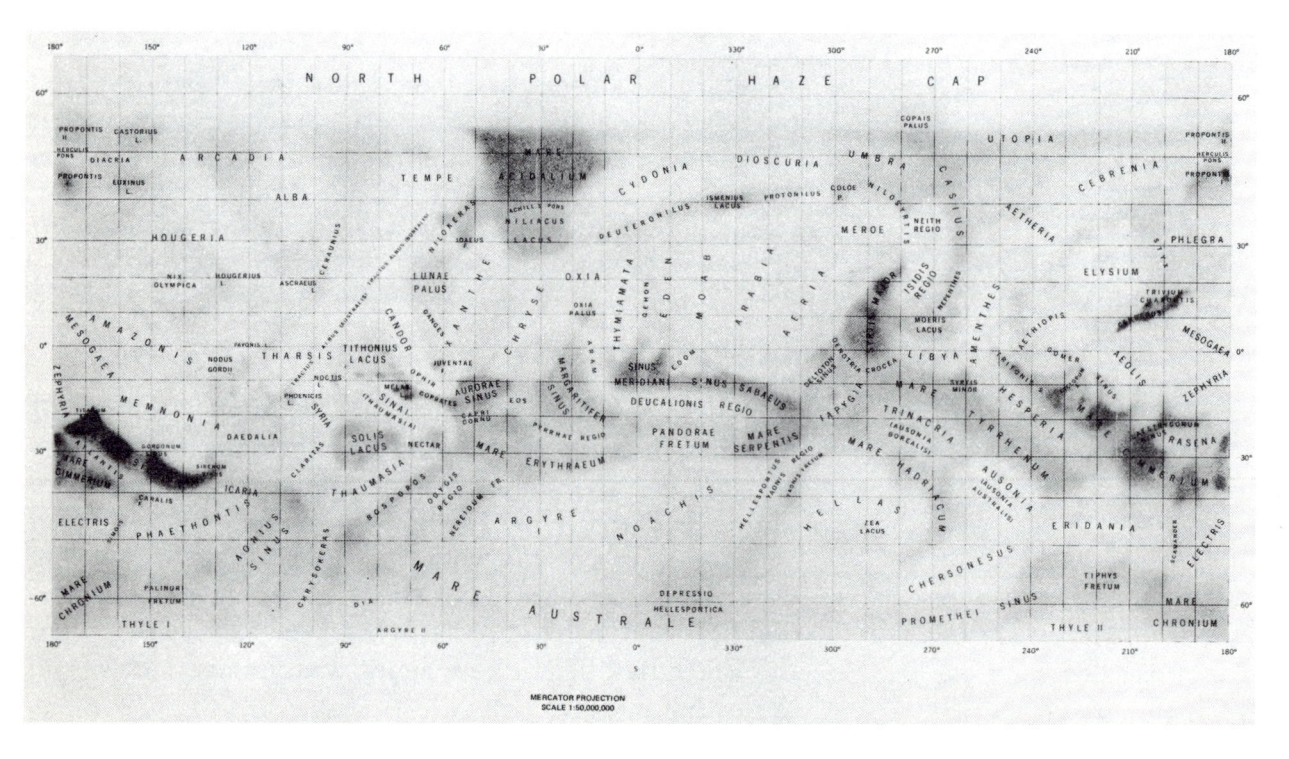

Amy Lowell, seen the *canali* as well. Lowell was a **Boston Brahmin**, educated and wealthy enough to do pretty much as he pleased. He liked astronomy, and was one of the first observers to locate a telescope in a location chosen specifically for the best seeing, choosing the isolation and cold dry air of Flagstaff, Arizona. He saw the *canali,* and they changed his life.

Schiaparelli died believing that there were seas on Mars. He spent years analyzing changes in the colors of the seas and comparing them to seasonal color changes in seas on Earth. Lowell in one stroke wiped out the seas. He saw canals running through them.

Year after year Lowell studied Mars. Having proved there were no seas, he went on to prove there were no large lakes. He found that water was very scarce on Mars, but he was certain that it had not always been so. There were ancient sea beds remaining, and there might even be rivers and small lakes.

The Lowell Observatory produced maps at great expense. An elaborate Mars globe was assembled. Surface details were mapped again and again, each one named or numbered, and the network of canals became more and more extensive and intricate. Lowell had no way to measure accurately the temperature, atmospheric pressure, or atmospheric composition of Mars. All he could do was study the *canali.* How could such an intricate net have come to exist?

For Lowell there was only one possible explanation. The *canali* were not natural channels but actual artifacts. They must have been built by intelligent beings to save their dying civilization. The seas were gone, the rivers were drying up, but still there was water in the polar ice caps. It melted in the Martian summer and the Martians pumped it through the canals to make their dying planet live a while longer.

That, too, was an appealing image. It was popularized not only by Lowell and by his European counterpart Camille Flammarion, but also by H. G. Wells in **The War of the Worlds** and by Edgar Rice Burroughs in his stories of Barsoom. John Carter's visits to Mars might be fantasy and the red and the green Martians a master storyteller's invention, but the background details could still be real.

The Old Mars took root and grew. For many the planet of ebbing life was as real as the interior of Africa. In fact, men and women of imagination became so certain that life existed on Mars that when essay prizes were offered to those who could prove the existence of life on other planets, Mars was generally excluded from the contest as being "too easy."

Every observed phenomenon on Mars was interpreted in the way most favorable to Martian biology. For example, if the temperatures were in doubt, as they came increasingly to be with more accurate determinations, the higher ones more conducive to life were always chosen.

Some people did have their doubts. A few highly skilled observers had always refused to accept the canals as real. Edward Emerson Barnard, the most keen-eyed of American astronomers with a score of comets to his credit, could never find them. George Ellery Hale reported from the Mount Wilson Observatory as early as 1909 that when he used the 60-inch reflector at low magnification he could see canals, but when he stepped up the magnification to 800 diameters the canals broke down into irregular blobs and splotches. Other reliable observers reported the same.

This skepticism made no impression on Lowell. When it was proved that the canals would have to be very wide in order to be visible from the earth, Lowell's followers found an easy explanation: the canals themselves were not visible, but we were seeing the bands of vegetation running alongside.

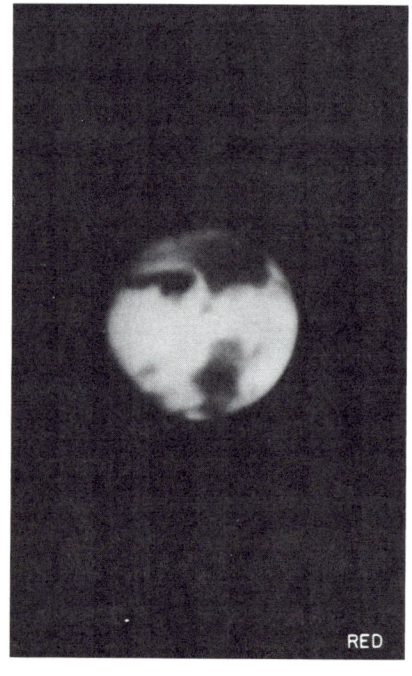

The best views of Mars from Earth, taken with the 200-inch telescope. (Mt. Wilson and Mt. Palomar Observatories)

A closeup look at Mars, taken by Mariner 9's cameras, shows a 700-kilometer-long sinuous valley. Is it a dry riverbed, or was it formed by the collapse of an underground lava tube?

The controversy over the canals actually continued until quite recently. When in 1951 Harold Spencer Jones, then the Astronomer Royal of London, published his book *Life on Other Worlds*, he devoted more than a dozen pages to the discussion, even though he was personally convinced that there were no canals. *A Space Traveller's Guide to Mars*, published five years later by I. M. Levitt (Director Emeritus of the Fels Planetarium in Philadelphia) did the same.

Gradually the arguments in favor of the canals died away. One of the last was an article in the March, 1956 issue of *Astounding Science Fiction,* in which Wells Alan Webb attempted to prove by a mathematical analysis of the visible networks that the canals were the work of intelligence and therefore they must exist. The networks, he maintained, were too complex to have arisen by chance.

He was perfectly right. The canals were indeed the product of intelligence—intelligence on the terrestrial end of the telescope. The human eye, aided by the legacy of Percival Lowell's persistence, tended to organize faint markings on the Martian surface and link them into straight lines and complex patterns that never existed.

There was a side effect of the canal controversy. The debate raged so bitterly, particularly in the period just prior to World War II, that according to Carl Sagan of Cornell University it affected all of astronomy: Many astronomers were so dismayed by the polemics of what seemed a profitless debate that there was a general exodus from planetary to stellar astronomy.

At last the canals were gone from serious scientific discussion. Lowell's Legacy, however, remained. The idea that there was life on Mars was firmly established and it outlived the canals that had been its principal evidence. Even though Jones, Levitt and the authoritative *Space Encyclopedia* of 1957 all held that there were no canals, they nevertheless flatly stated that there was great evidence for the existence of vegetation on Mars, although, sadly, it was "now known that conditions were unfavorable to large animal life."

The canals had done other work. It would have been impossible to pump large quantities of water from the poles to the equator of Mars if there were high mountains on the red planet. Therefore, Lowell had reasoned, there were no high mountains. Mars was a land of rolling plains and low hills. This too outlived the canals which had generated the theory. Science fiction stories almost universally assumed gentle Martian terrain. Levitt could state that "the surface of Mars must be monotonously flat." There were observations suggesting otherwise, but they were ignored.

The canals had suggested life. They were discredited, but the Martian life remained. It was observed that each summer a "wave of darkening" spread from the melting pole to the Martian equator. Although it moved quite rapidly, several kilometers an hour in some times and places, it was almost universally ascribed to vegetation "awakened" by the coming of moisture. A few hardy souls proposed physical causes such as dust storms, but nevertheless the *Space Encyclopedia* could state "there are obvious disadvantages to these theories, and it is fair to conclude that the existence of lowly vegetation on Mars is very probable."

Mars's atmosphere did not escape misinterpretation. Spectrographic data showed absorption features in the infrared region that were instantly attributed to chlorophyll. Even when the features were later recognized to be heavy water (HDO: water with a deuterium atom replacing one of the hydrogen atoms) there was a brief attempt to postulate that Mars was rich in heavy water instead of accepting that the substance was in the earth's atmosphere and had not been removed from the spectra of Mars. As Bruce Murray of the California Institute of Technology points out, a Mars rich in heavy water would still be attractive because it could "only have come about if large quantities of normal water had once been present on the surface of Mars over very long periods of time—that is, under very earth-like conditions."

That was not the extent of the contrary evidence. From time to time there were observations that indicated that the Martian atmosphere was not 10 percent as dense as the Earth's, but was actually less than 1 percent as dense. They were ignored. It was easier to believe in Martian air thick enough for human beings to walk in without the need of pressure suits—and of course life would be unlikely in conditions at 1/100th the atmosphere pressure of earth, whereas it not only could but did exist on earth at 1/10th atmospheric pressure.

"Mars jars" became a scientific fad during the late 50's and early 60's: scientific simulations of what we thought was the atmosphere of Mars. Bacteria, and even complex plants such as pumpkins, were put in the Mars jars where they happily grew. The possibility of life on Mars was "proved." If Earth plants needed only some water and a minimum of air, certainly Mars plants would have no difficulties. The problem, of course, was that the Mars jars did not contain a Martian atmosphere. The air inside the jars was too thick and possessed too much water, and sometimes it was composed of the wrong elements as well.

Lowell's Legacy has been far reaching, and we have not

seen the end of its influence. Our very space program was structured by it. As late as October, 1964 the Space Science Board of the National Academy of Sciences stated:

> The primary goal of the national space program in the exploration of the planets is Mars: it is one of the nearer planets (and hence relatively accessible); as a planet, its biological, physical, chemical, geophysical, and geological properties are at least as interesting as those of any of the other planets; of even greater significance and excitement to mankind, it affords the more likely prospect of bearing life.

Yet at that time there was not one single observation of Martian organic constituents there, much less biological ones. There was then and is now no evidence for life on

The south polar cap of Mars, photographed by Mariner 9 *during its fifth orbit of the planet, in 1971.*

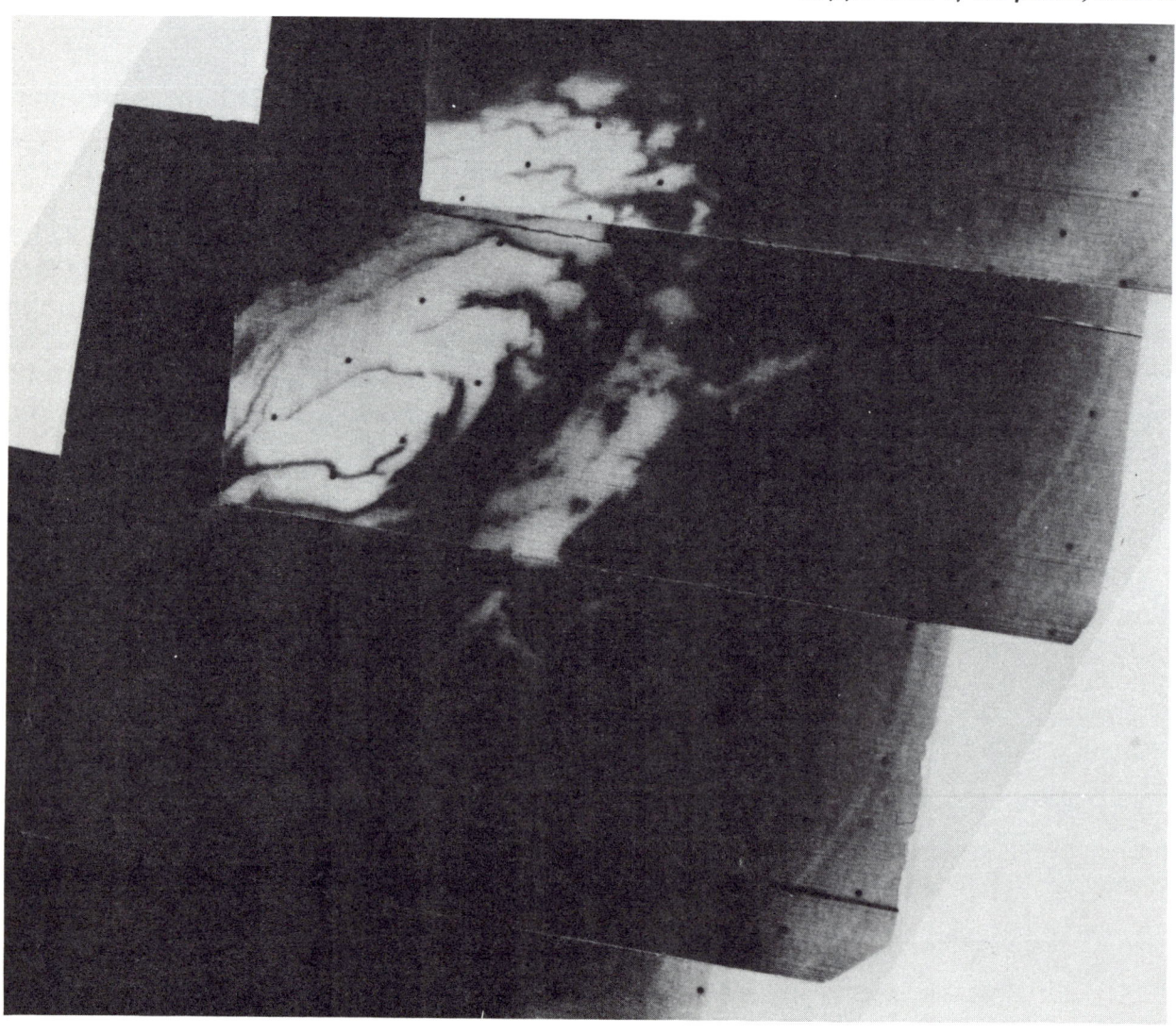

The "Grand Canyon of Mars," Valles Marineris, is almost as long as the United States is wide. The three craters lined up from just above Seattle to slightly west of San Francisco are North, Middle, and South Spots, in the region of Tharsis.

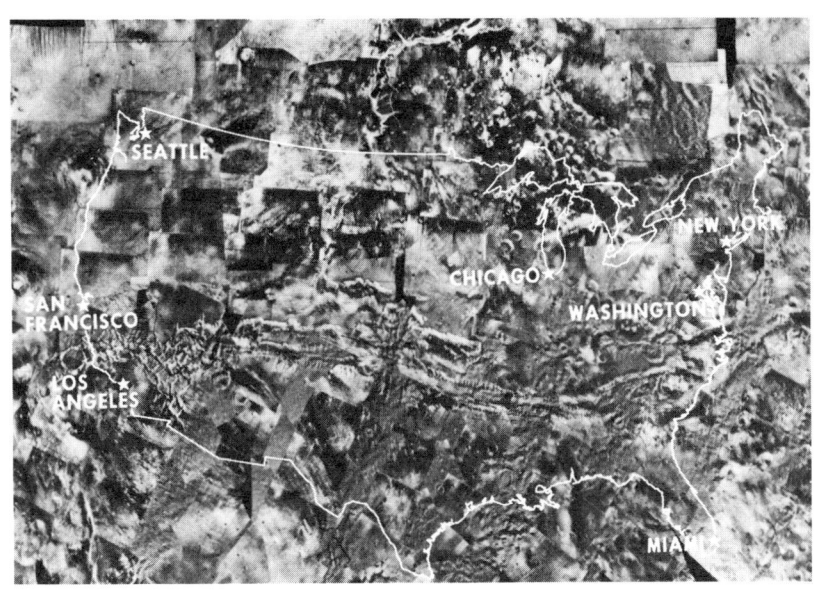

Mars. Even so we have sent two space probes designed to detect Martian life, and we are doing this before we have—indeed, in lieu of getting—good measurements and maps of the Martian poles. The analysis of the polar ice caps has varied with the prevailing climate of opinion. Although observers at the turn of the century believed that the ice caps were composed mostly, if not entirely, of solid carbon dioxide or dry ice, that hypothesis was discarded. A number of experiments were designed to "demonstrate" that the ice caps were made of water. For example, Levitt pointed out that because of the low pressure on Mars "dry ice cannot continuously exist at a temperature higher than -230 degrees Fahrenheit, and this rules out this material, because the temperatures on Mars are far above this figure." There was, of course, strong evidence that the temperature at the Martian poles was much lower than we then supposed, but it was conveniently ignored.

As contrary evidence poured in it was just as swiftly explained away. The Old Mars retained its foothold with an almost incredible tenacity. The Old Mars was a dying world that had once possessed a thick earthlike atmosphere and still possessed remnants of it, a world that had once had flashing seas and running rivers but now held only tiny amounts of water locked in the polar ice caps to be released each summer when it spread life across arid deserts. According to the *Space Encyclopedia,* "winds on Mars are very moderate, and appear incapable of raising giant dust storms; nor is vulcanism likely." It was a world that had once been home to Martians who built canals, the grandest survival effort in the solar system. It was a world where men would one day stroll along the Grand Canal and see in the distance fragile towers built by a race long dead.

It was the world of Lowell's Legacy. Life could exist on Mars, and since it could exist it did exist. And because it did exist, Mars was by definition a world where life *could* exist. It must.

THE LUNAR MARS

Mariner 2, the first interplanetary probe launched by NASA, went to Venus, but Mars was the space program's first love. The Apollo program that eventually placed man on the moon made early unmanned interplanetary exploration almost a sideshow, but Mars nonetheless received its share of attention. In fact, it received almost too much attention. Had we launched the early Mars lander some wanted, it would have failed. It was designed for a much thicker atmosphere than we now know Mars to have.

Mariner 4, the first successful Mars flyby, was designed and built by California Institute of Technology's Jet Propulsion Laboratories and launched on 28 November, 1964. It performed flawlessly, and on 14 July of the next year it passed within 9900 kilometers of Mars. The probe sent back 21 pictures, the first ever taken of the surface of another planet.

At JPL we waited breathlessly. At eight bits per second the pictures took a long time to arrive. Each bit was a single point of light or dark, and it took thousand of points to make up a single picture. Finally the photographs of Mars, undistorted by the Earth's atmosphere and taken from thousands rather than millions of kilometers above the Martian surface, were shown to newsmen.

There were no canals. There were no rolling plains. There were no oases, or cities, or Martians. There were only craters.

Mars looked exactly like the Moon.

Richard C. Hoagland suggests in the December, 1974 issue of *Analog Science Fiction* that when those pictures arrived, public interest in the space program began to die. If there were no Martians and never had been, no Grand Canal and no hurtling moons of Barsoom; if indeed as JPL announced, "Mars is dead, cratered and battered like the Moon, and thus it has been for the life of the Solar System," then the reason for interplanetary exploration died with the Martians who had never been.

The other experimental results were even less encouraging. There was no magnetic field around Mars. The atmosphere was assuredly less than 1 percent the density of the Earth's. The surface of Mars had no water; it was a planet of dry blowing dust.

Momentum sent more spacecraft to Mars, but it seemed that only the planetary investigators cared any more. Something had gone that could never be recovered. The sense of wonder about Mars had died in all but the most scientifically determined hearts.

The next two Mars flyby probes, *Mariner 6* and *Mariner 7*, gave the planet even worse press. They arrived at Mars on 30 July and 3 August, 1969, respectively, only days after the triumphant moon landing of *Apollo 11* and were far less spectacular by comparison; furthermore they confirmed the ill tidings of *Mariner 4*. The polar ice caps were again found to be solid carbon dioxide. Once again radiometric measurements showed no trace of water and little atmosphere. Once again the cameras photographed craters. As if all that were not bad enough, there was not even any heat inside the craters. They were undoubtedly primordial. They showed no evidence of ever having been eroded, inescapably implying that there never had been a period during which Mars had had flashing seas and running rivers.

And yet, the craters were not *quite* like those on the Moon. Clearly some very interesting geophysics had been a part of Mars' past, and we did not understand them. Two views immediately formed about Mars and its place in planetary exploration, and they have lasted to this day. Murray and his colleagues at Cal Tech said that Mars was more like the Moon than anything else, but even without life it was an intriguing place and well worth study. They cautioned, however, that it should not be investigated to the exclusion of the other planets. We ought to take a first look everywhere rather than concentrate too much on any one place at the beginning.

Sagan and his colleagues at Cornell University held a different view. Mars was more like the Earth than the Moon in many ways; not only did we not understand what had happened to Mars but perhaps it was even far more interesting than we imagined. Although Sagan had developed the nonbiological theory to account for the wave of darkening that had been cited as primary evidence for biological processes on Mars, he had not rejected the concept that Mars might still yet possess life—even though after *Mariner 6* and *Mariner 7* it was difficult to see where the life might be or what kind of life could even survive in that bleak and hostile world.

Because Murray and the Cal Tech team were interested in Mars as a planet with a fascinating geophysical history, they wanted closeup looks at the hitherto ignored polar ice caps and as much of an overall view of the planet as possible. Therefore they proposed a polar orbit for one of the two upcoming Mars orbiters.

Sagan and his team, on the other hand, were principally interested in exploring the kinds of phenomena that were visible from the earth, such as the areas of light and dark, and the wave of darkening. They were primarily interested in pictures of the equatorial and medium latitude regions. They proposed an orbit inclined only 50 degrees with respect to the Martian equator.

There was no conflict as long as two orbiters were planned. When *Mariner 8* was launched, however, it ended up not in orbit around Mars, but in the Atlantic Ocean. A decision now had to be made. *Mariner 9* might be the only opportunity in this generation for a close and detailed look at Mars. To be sure, the ambitious *Viking* lander was on the drawing board, but it might not go. A polar orbit would yield more information about the planet in general, argued the Cal Tech group. A low-inclination orbit, on the other hand, would give a more detailed picture of the equatorial region, contended the Cornell group; it was believed that *Viking* should land there since presumably that would be the best place to find life. Nonsense, replied the geophysicists. Surely it is obvious that either there is no life on Mars, or if there is it is so strange that we cannot make any meaningful predictions about where it can be found.

The decision was finally made: *Mariner 9* would go into a 65-degree orbit, the mean between that wanted by the Cornell astronomers and the polar orbit desired by the Cal Tech geologists. It satisfied neither.

It was one more bequest of Lowell's Legacy.

Mariner 9 was one of the most sophisticated objects ever built by man. It crossed 100 million miles of vacuum

Sinuous channel appears strikingly similar to terrestrial riverbeds.

Mariner 9, *the first spacecraft to orbit a planet beyond the Earth-Moon system.*

to become the first manmade satellite to go into orbit around another planet, and sent back data and pictures at a rate of 16,000 bits per second, using less power than a 20-watt lightbulb. The transmitter obtained its power from solar receptors placed on the spacecraft's cruciform "wings". *Mariner 9* used cold stored gas to maintain its attitude in space, first to keep the solar panels directed toward the sun, and then turning to aim the antenna toward the Earth. The gas was used up after nearly a year of operation.

Although the craft remains ready to work to this day, it cannot turn toward the Sun to get the power it needs. Ironically, there is gas aboard, in the engine that was used to put the craft into orbit around Mars. It would have cost thirty thousand dollars, however, to modify the spacecraft so that it could tap that gas for attitude control. The modification was proposed, but was rejected for budget reasons. The total Mariner program cost about 100 million dollars.

Mariner 9 carried two television cameras, one a wide angle, low-resolution camera for mapping the surface of the planet, the other a narrow-angle telephoto system for detailed observation. In addition, *Mariner 9* carried a variety of scientific instruments.

An ultraviolet spectroscope measured atmospheric pressures and identified gases in the upper atmosphere.

An infrared radiometer measured temperatures on the surface. An instrument called IRIS, for *InfraRed Interferometer Spectrometer* was included as well. It was a clever device that gave not only measurements of atmospheric temperatures and pressures near the surface of Mars, but also yielded information about the composition of the lower atmosphere and some clues as to the composition of Mars's soil. One of the major tasks assigned to the IRIS was to settle forever the debate about the composition of the polar ice caps: were they dry ice, water ice or both?

The spacecraft itself was also used as an "experiment." Since its orbit could be precisely determined, perturbations in the orbit as the craft circled the planet would indicate local variations in the gravity and thus in the surface and mass of Mars. Moreover, when *Mariner 9* passed behind Mars for a short time during each orbit, the occultation of the radio signal by the planet yielded further information about atmospheric temperatures and pressures.

The data for the photographs were sent back in digital form, mere strings of numbers that had to be fed into JPL's big computer before an image could be built up. In doing so the computer had the option of enhancing the data, that is, emphasizing and increasing the contrast so that the resulting pictures were as good as or better than the view a live astronaut would have had from the spacecraft.

On 13 November 1971, at precisely 4:21 PM Pacific Standard Time, the engine aboard *Mariner 9* was scheduled to activate, ending the craft's six months of silent coasting through space. Fifteen minutes later the probe would be in orbit around Mars; six minutes and 20 seconds after *that* the controllers at JPL would know whether or not their mission had succeeded.

The spacecraft operated perfectly. Unfortunately, Mars refused to cooperate. During the time that *Mariner 9* had been on the way to the red planet, the largest dust storm of the century had appeared. It covered the entire surface. Nothing remotely like it had ever been seen and it showed no signs of going away.

The original Mariner mission plan had called for a series of pictures of the entire planet to be taken as the spacecraft approached. As *Mariner 9* drew closer, the resolution of its photographs would improve, thus not only showing greater detail, but giving automatic

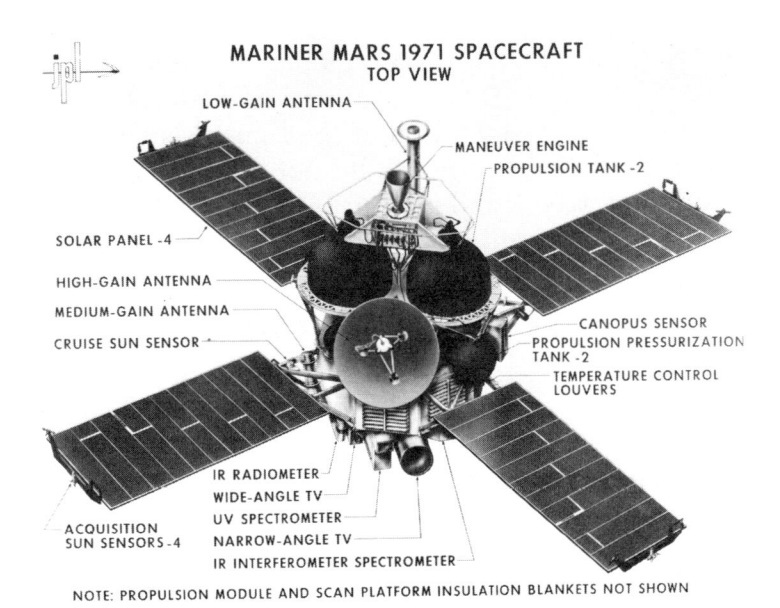

Schematic drawings of the Mariner 9 *spacecraft.*

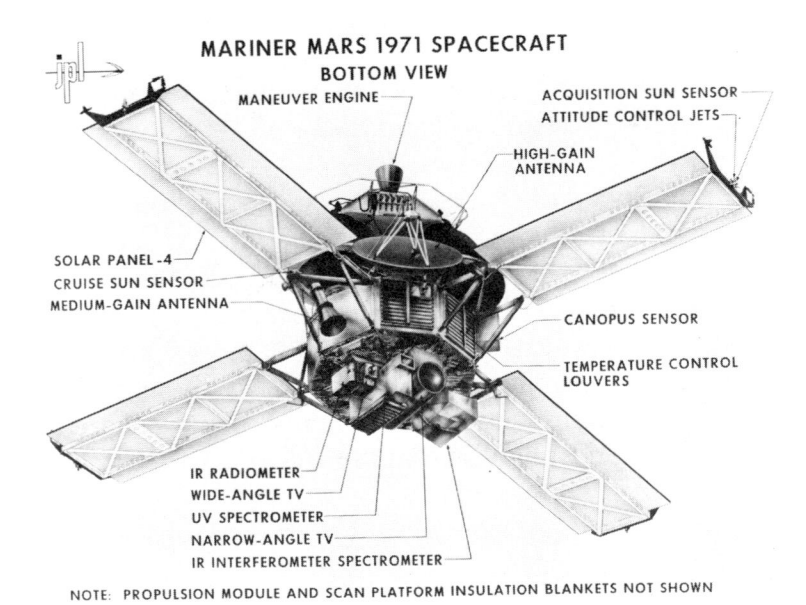

interpretations. Instead, they showed nothing. The craft faithfully transmitted picture after picture of a uniform grey field. A few spots jutted out from the featureless mass, but that was all. Even the computer enhancement techniques did not improve the results.

"We spent hundreds of millions of bucks for this?" one reporter asked.

A cameraman muttered: "First it was sharp pix of bedsheets. Now they're giving us enhanced bedsheets."

"A marvelous opportunity," the meteorologists exclaimed. "A unique opportunity to learn about Martian weather." The excited voices and wide smiles seemed a bit forced.

Unlike the cratered world of the Lunar Mars, the New Mars revealed by *Mariner 9* did not dramatically spring forth. It took weeks and months for the story to come out and by that time public interest in Mars had evaporated—just before the most exciting discoveries since Lowell thought he saw canals.

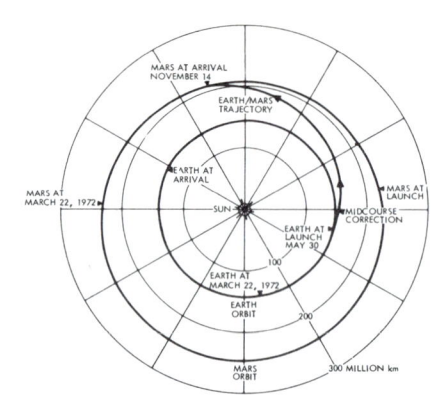

Mariner 9's *six-month-long path to Mars. Although designed to function for 90 days once in orbit around Mars, Mariner 9* continued to perform satisfactorily for 349 days in orbit around the Red Planet.

THE NEW MARS

When the dust storm finally cleared, Mars was revealed as a world in its own right, like neither the Earth nor the Moon, with all the variety one would expect from an entirely different planet. Far from being a dead and cratered world, Mars proved to be a planet of such varied terrain and structure that we still have no good theories to explain much of what we have seen.

The Many Faces of Mars Mars possesses four types of features that are the result of violent volcanism—but for reasons that we do not yet understand, all of them are confined to the northern hemisphere. The southern hemisphere is for the most part densely cratered, superficially resembling the lunar highlands. By sheer bad luck all the pictures taken by *Mariners 4, 6* and *7* were in the cratered hemisphere so that the volcanic structures of Mars were entirely unsuspected.

The most dramatic volcano is Olympus Mons. It was previously known from terrestrial observations as Nix Olympica, "The Snows of Olympus," because it periodically develops bright areas suggesting snow. Its true nature was a total surprise. Olympus Mons is a gigantic volcanic caldera, containing more mass than all of the Hawaiian Islands together. It rises nearly 30 kilometers above the surrounding plains. This is nearly half again the distance from the depths of the Mariana Trench, the Earth's lowest sea trough, to the top of Mount Everest. Thin vapors trail from its lips, causing some investigators to think it may still be active and outgassing.

A line of three volcanoes, smaller than Olympus Mons but as large as Earth's greatest, lies along the nearby Tharsis Ridge. Like Olympus Mons, the Tharsis group are shield volcanoes or lava domes, with gently sloping sides built up from repeated outpourings of lava. Their centers have collapsed to leave ring-shaped calderas that superficially resemble lunar craters. Because they were three of the marks visible on the bedsheet photographs during the great dust storm, they were dubbed North Spot, Middle Spot and South Spot. Their real names, however, are Ascraeus Mons, Pavonis Mons and Arsia Mons. Arsia Mons alone rises some 17 kilometers above the floor of the Amazonis basin to the west.

The Martian shield volcanoes, the first type of volcanic feature, are substantially larger than their terrestrial counterparts. The Olympus Mons complex is 600 kilometers across, and according to Volume II of the *Mariner Final Report*, "there is some evidence that its size has been reduced significantly by erosion." Ascraeus Mons, Pavonis Mons and Arsia Mons are each 400 kilometers across. The shields all have the same general structure; they are all roughly circular, they have central depressions at their summits, concentric features, and steep escarpments. There is little doubt that they were built up by processes similar to those that formed such terrestrial volcanoes as Thera in the Aegean Sea.

The Tharsis region also has smaller volcanoes with very steep sides. These domes, the second type of volcanic feature, may be young shield volcanoes or shields that have had only small lava flows, or they may have been formed by entirely different processes.

In the Elysium and Tharsis regions there are several volcanic specimens of a third variety. They have scalloped outlines and the rims of their craters are smooth

The heavy line on this map shows how Mars is divided into two distinct halves, as far as physical features of the surface are concerned. The southern hemisphere seems to be older and much more heavily cratered, while the northern hemisphere is less cratered but studded with volcanoes, some of which may still be active.

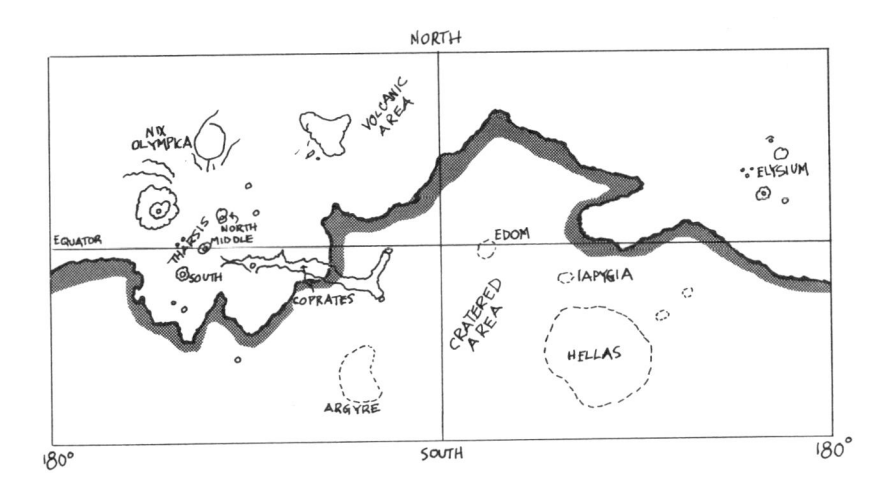

49

surfaces terminating against the surrounding terrain without any sharp break in slope. Their general shape is distinctly circular, but they do not seem to be impact craters.

The final class of volcanic feature comprises the large plains that are similar to the lunar maria. Some of the plains overlay terrain pocked with craters, whereas on other parts of the planet the plains themselves have clearly been repeatedly cratered after they were formed. Michael Carr of the U.S. Geological Survey speculates that volcanism was an active process over the entire planet but that the hemisphere resembling the lunar highlands was afterwards bombarded by large objects.

He also concludes that Mars has never experienced the crustal activity known as plate tectonics that has so shaped structures on the Earth: the movement of large areas of the crust because of convection currents within core and mantle. He notes that if Olympus Mons was built up at the same rate as the Hawaiian Emperor Chain of islands, it would have taken at least 130 million years to construct the visible shield.

Since we are not seeing all of Olympus Mons (some of the shield has been destroyed by whatever formed the escarpment around it, and part has been hidden because the ground has subsided under that great weight of mountain), Olympus Mons could have been accumulating for up to a billion years. All this implies that the crust on Mars is stationary with respect to the mantle, because as Carr says, "astonishing rates of volcanism would be required to build Nix Olympica if the Martian crustal plates were moving with respect to the magma source." Of course, Mars has astonished us before.

It was hoped that *Mariner 9* might find active volcanism, and "hot spots" were reported to the press. However, none of these were due to current volcanic activity. They turned out to be dark areas which absorb more sunlight, or the west sides of slopes at sunset. Although no lava flows were found, this could have been due to instrument limits. The radiometer could only examine wide areas, and thus a small hot spot would vanish in the general average temperature of region. Moreover, the investigators point out that "even if Mars had twice the volcanic activity of Earth, the probability of observing such [a volcanic feature] would be very small."

One of the most spectacular finds of *Mariner 9* was a great gash across Mars. Named Valles Marineris or Mariner Valley after the spacecraft that discovered it, it is sometimes referred to as the Martian Grand Canyon, and is longer than the U. S. is wide. In places the valley floor is six kilometers below the rim. Valles Marineris is part of a complex of canyon lands in the Tithonis Lacus-Coprates region.

Planetary scientists feel that "canyon" is not an appropriate name for this complex of features. Some of the canyons are depressions closed at both ends, and others, although joined together, do not compose an integrated trunk and tributary system, nor are their floors smoothly graded. Thus the term "troughed terrain" is preferred, since this does not imply how these spectacular features were created.

There are dozens of such troughs in the equatorial region. They range up to 200 kilometers wide and several hundred kilometers long. The walls are steep with their upper parts scalloped into U-shaped chutes descending from the brink and giving way to smooth slopes below. The walls do not appear to be layered or stratified, but this may be due to limits of camera resolution; layers thinner than tens of meters would not be seen. Many of the troughs have blunt abrupt ends.

There is no generally accepted theory of what caused these astonishing features. There is nothing like them on Earth. It is generally thought that the troughs were caused by some kind of fracturing of the Martian crust, since what the geologists call graben on Earth—depressions caused by sinking of the land between two parallel faults—are clearly seen in the region, and some run into the giant troughs. This has led Robert B. Sharp of Cal Tech's Planetary Sciences Division to suggest that the great troughs were formed from chains of pits between two very large parallel fractures of the Martian surface.

According to this theory the pits grew larger from Marsquakes, "possibly aided by wall recession through sapping" but largely through further collapse, until the pits merged. This would account for the scalloped walls. Sharp, however, is quick to say that this is speculation. Even more speculative is his theory that wall recession resulted from deterioration of exposed ground ice which evaporated or melted. If the subsurface were permafrost exposed by the initial pit collapses beginning the process, it would quickly melt or evaporate, and what remained might be finegrained particles. The melting would undercut the walls, more material would collapse, and more permafrost would be exposed.

However, this would dump huge quantities of rock debris onto the trough floors. The troughs are closed at both ends, so that material could not have been eliminated by flowing water—and no one can explain what has happened to it. The enormous volumes involved make eolian (wind-carried) erosion extremely unlikely: all of the rocks would have to be ground to a fine powder, and no one knows how that could be accomplished.

It has also been suggested that most of the debris was ice that has melted away, but Sharp calculates that would have required more than a million cubic kilometers of water. If the entire Martian surface were underlain by the several kilometers of ice required to form the troughs by evaporation, Mars would have nearly a quarter as much water as Earth. Since the surface area of Mars is about a quarter that of Earth and we observe almost no water, this seems unlikely. It is therefore generally believed that somehow the debris must have gone downward into the planet's interior.

Some theorists wonder if Valles Marineris is not much like the Great African Rift on Earth, implying the beginning of plate tectonic activity on Mars. Murray, who elsewhere has written about the necessity for scientists to be able to live with uncertainty and keep several alternate theories in mind at all times, argues that Mars may have only recently begun to heat up and develop volcanism. In some bizarre fashion the debris from the troughs may have been carried away to the newly formed volcanoes and spewed up as lava.

It requires a pretty far-out theory to make Valles Marineris similar to the terrestrial Great African Rift, because nowhere on Mars is there evidence of young mountains being thrown up by the spreading apart of tectonic plates as there is on the Earth. In fact, as Murray himself is quick to say, the problem of explaining the Martian troughs is not one of choosing between competing theories: there is *no* theory that adequately explains what is known about them.

Spectacular as Valles Marineris is, another and very different kind of troughed terrain has attracted even more

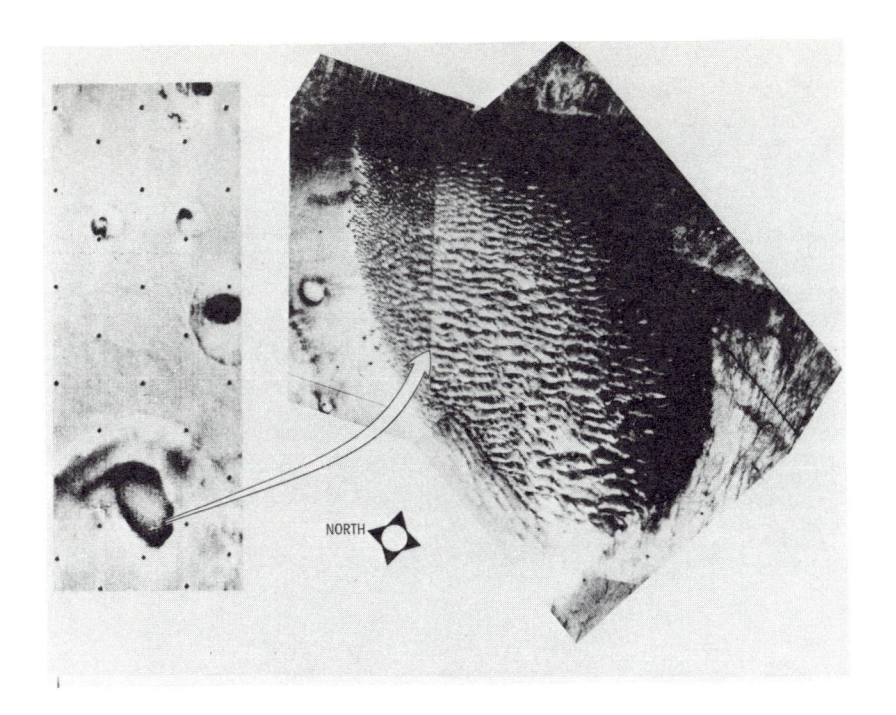

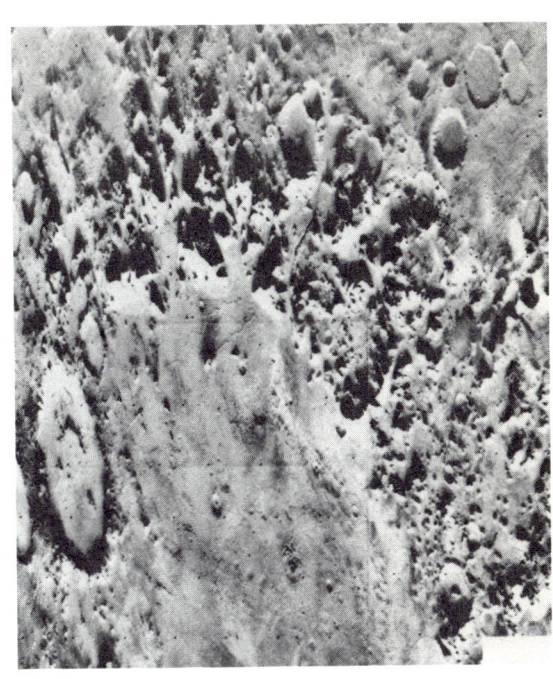

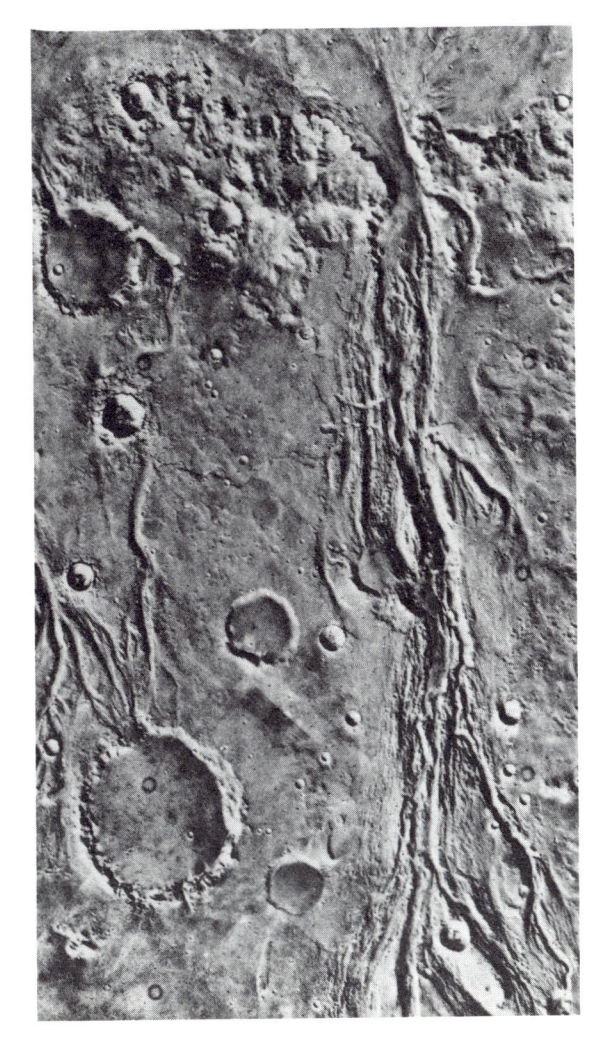

The varied face of Mars. Sand dunes lie at the bottom of a Martian crater (upper left), *while* Viking *orbiter photograph* (upper right) *shows plain of Argyre. "Fluvial" landscape in Chryse Planetia area* (lower left) *suggests extensive flooding in past era. West end of Valles Marineris, as seen by the* Viking 1 *orbiter from 4300 kilometers (2700 miles) range. The canyons run east-west and are about 60 kilometers (37 miles) wide.* (lower right)

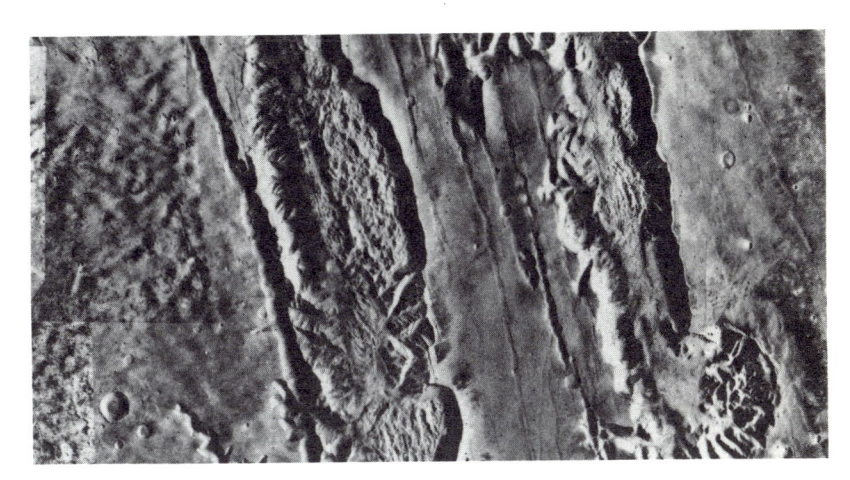

53

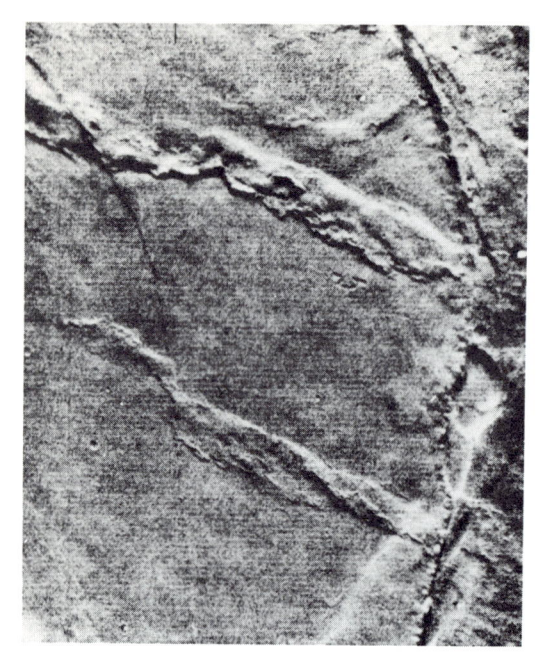

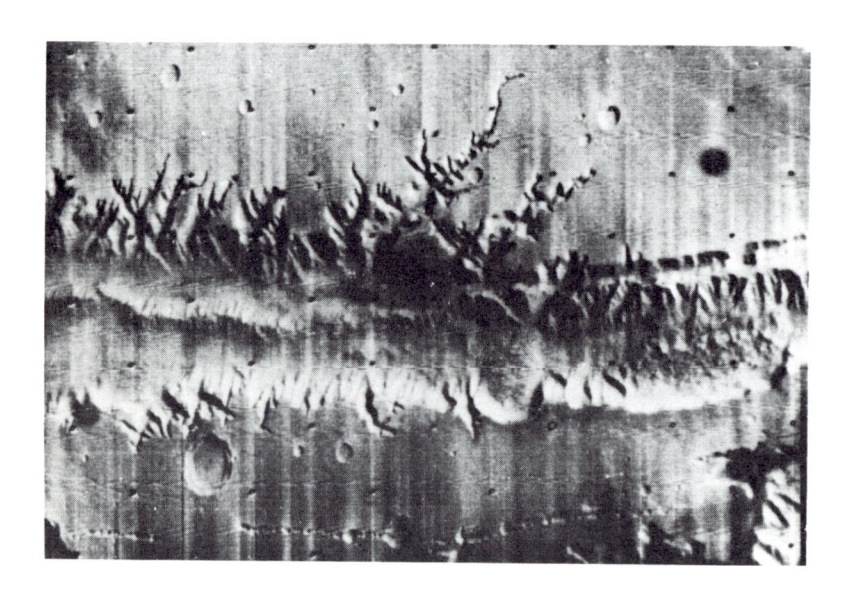

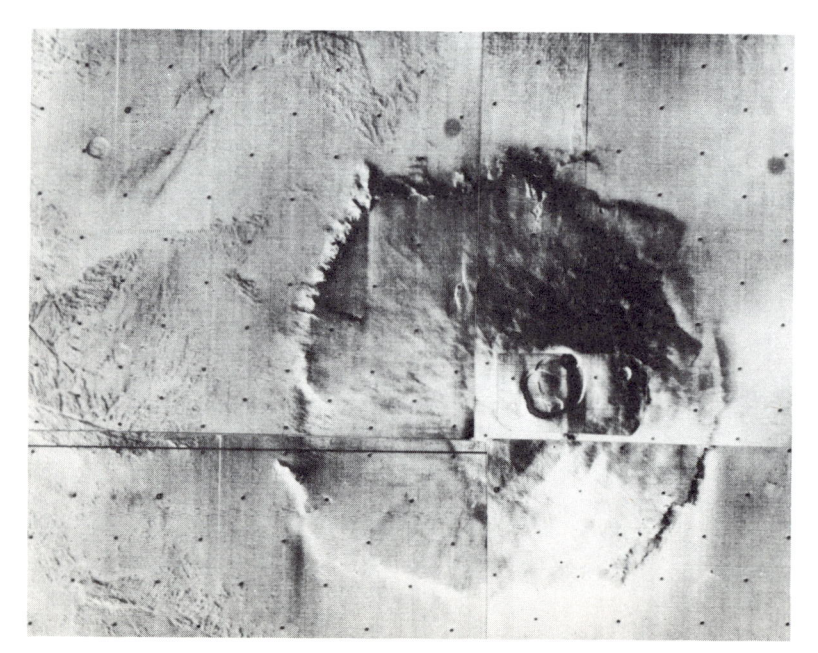

Mariner 9 *views of Mars's surface. The crater (upper left) near Pavonis Lacus appears to have been caused by the collapse of a volcanic caldera, rather than impact of a meteor. Ridges shown in closeup photograph are similar to lunar ridges caused by lava flow. Canyon system (center left) in Tithonius Lacus region has many tree-like tributaries, like terrestrial stream systems. Olympus Mons (lower left) is the highest mountain in the solar system, towering 30 kilometers tall (Mt. Everest is 8.81 kilometers high). This giant volcano may still be active. Another view of Valles Marineris (lower right) showing a profile of its cross-section. This gigantic valley system is closed at both ends, and unlike any geological feature of Earth.*

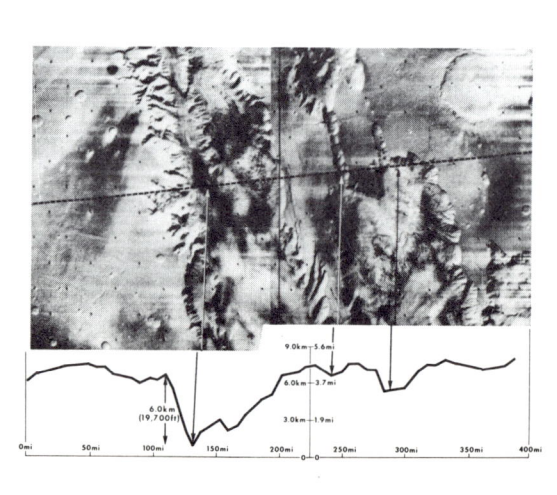

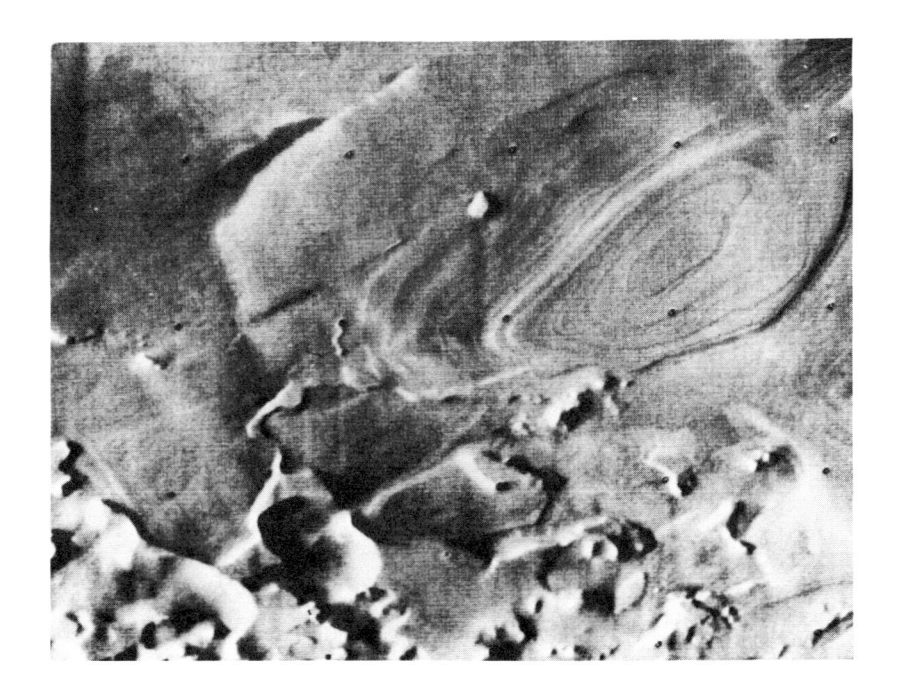

Poles, spots and streams on Mars. The Martian south pole (top left) shows terraced terrain, possibly caused by successive layers of ice freezing and then thawing incompletely. North Pole (right) ice cap also shows detailed structure. Arsia Mons (bottom) was dubbed "South Spot" during the first days of Mariner 9 mission, when all of Mars was covered by a planet-wide dust storm, except for a few mountainous "spots" poking up above the dust clouds.

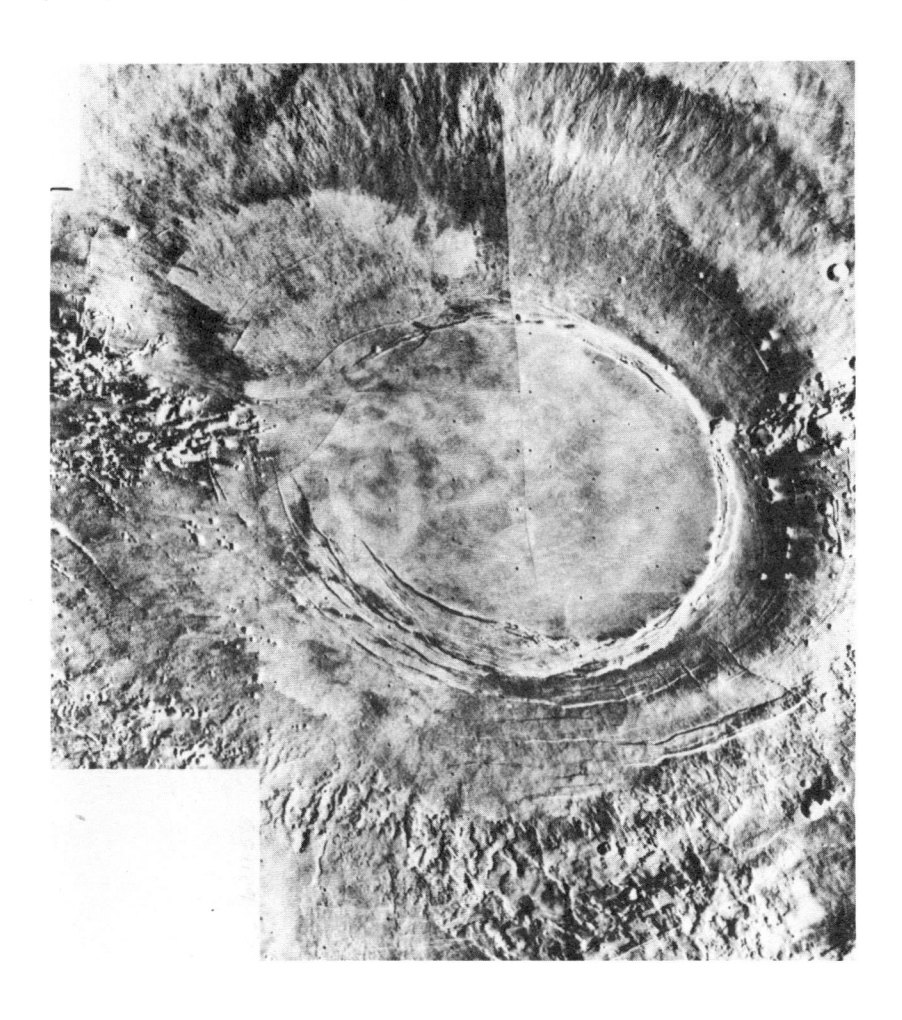

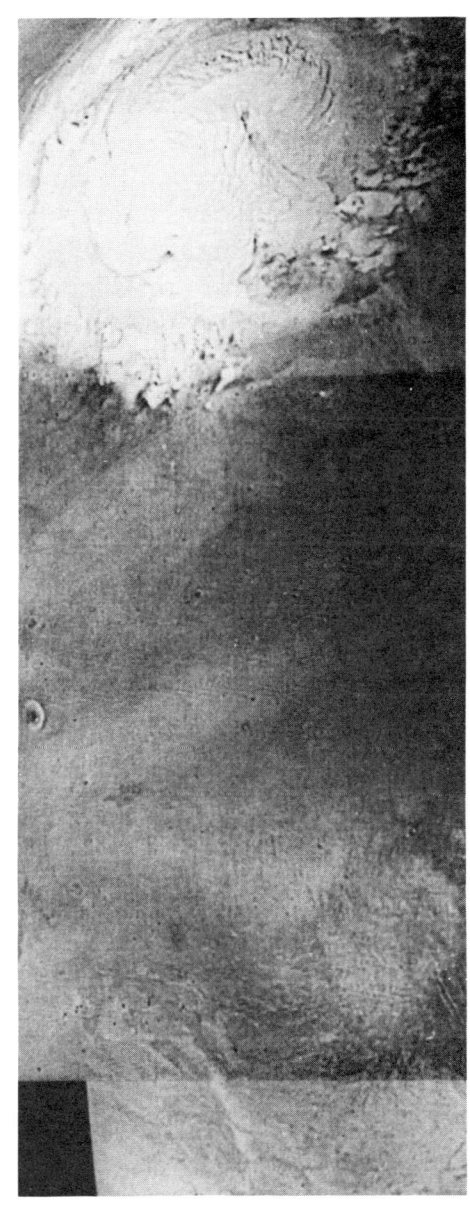

attention from both scientists and the popular press. These appear to be nothing less than river valleys complete with a developed system of tributaries and definite outlets.

These channels—some go ahead and call them *canali*—"deeply penetrate the surrounding uplands and debouch in tributary fashion into areas of smooth lowlands." They have steep sides and flat, smooth floors. If observed on Earth no one would hesitate to call them dry river beds, and in fact most planetologists agree that they were formed by some kind of flowing liquid. The problem is, what was it, where did it come from, and where has it gone?

The river-bed or *arroyo* troughs have generated spectacular theories about the geophysical history of Mars. Since these theories directly relate to the question of whether there is life on Mars they will be discussed in a later section.

In addition to Earth-like features such as volcanoes and troughs, Mars has large basins similar to the lunar maria. The largest is Hellas, a great flat area some 1,600 kilometers across. Its floor is one of the lowest areas on Mars. Other basins are Argyre, Libya, and Edom. All are generally circular, with walls composed of material different from the fill or floor.

They are generally believed to have been formed by the impact of very large meteorites, as were the lunar maria, although various features of the lunar basins are sometimes missing in the Martian examples. The most obvious difference is that the older Martian basins do not have complete rings of smaller craters formed by secondary impact of material ejected from the basin floor when the meteorite struck. Traces of the secondary rings have been found. The rest have been obliterated by repeated impact of smaller meteoroids. Thus the Martian basins are quite old. They are found only in the cratered southern hemisphere. Any that may have formed in the young volcanic hemisphere have evidently been covered by subsequent lava flow.

The basins, particularly Hellas, change color periodically. Sagan explains the changes as the result of local dust storms. Hellas is probably a vast bowl of dust, usually serving as a sink for windblown dust but in planetwide dust storms it may be a source as well. Sagan hypothesizes that the surface of Mars is dark, and the wind-blown dust

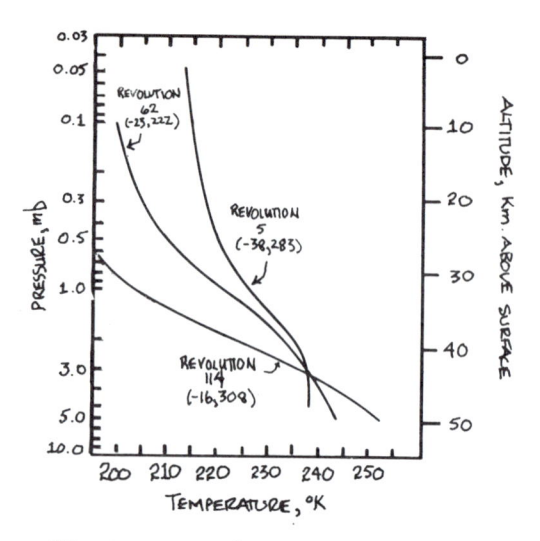

Martian atmospheric temperatures and pressure at three locations and times, measured by Mariner 9's *instruments. The Revolution 5 measurements, taken during the Great Dust Storm, show that the atmosphere behaved very differently while the storm was in progress.*

is light colored, so when the dust is deposited on the dark Martian surface it causes the color changes visible from Earth.

The same explanation would account for the wave of darkening each spring. As the polar caps melt, carbon dioxide blows from the pole to the equator, lifting the dust and revealing more of the darker surface beneath. Observations by *Mariner 9* make it nearly certain that this theory is correct.

In addition to the dust-filled basins there are large plains of what appear to be sand dunes. One field is 70 kilometers across and lies in the bottom of a crater. The dunes are spaced one or two kilometers apart, and the field is very similar to the Kelso Dune area of the Mojave Desert.

It is thought that wind erosion contributes to the physical form of some of the large trough features, particularly of Valles Marineris. That great canyon is so long that the ends are not at the same temperature: the east end can warm up each Martian morning while the west end is still in darkness. A wind blowing down the canyon over a period of millions of years could scour and etch features as well as grind up material that would be removed in planet-wide dust storms.

Mars's atmosphere is so thin that an observer on the surface would see space overhead as black, with stars shining brightly even in the daytime. Closer to the horizon the sky would be blue and the stars would not be apparent. Nearly every type of cloud seen on Earth has also been observed on Mars. The clouds are probably both condensed carbon dioxide and water vapor. Massive frontal systems were observed, as were temperature inversions at different altitudes and a number of other earth-like meteorological phenomena.

The observed Martian atmosphere is at least 90 percent carbon dioxide (CO_2), with traces of molecular oxygen (O_2) and other gases. However, there is recent indirect evidence that a large part, perhaps as much as 35 percent, of the Martian atmosphere may be the inert gas argon. Argon would not be detected by Mariner's instruments. The presence of clouds shows that some areas are saturated with water vapor, but the temperatures are so low that the actual amount of water is extremely small. If all of the water vapor were to precipitate and fall as rain, it would cover the surface under it with about 10 microns of liquid. (One micron is approximately 1/25,000th inch.)

The atmospheric pressure at the surface of Mars varies from a maximum of 0.9 percent that of the Earth at the lowest areas to 0.3 percent on the highest plains and canyon lips. Those pressures are equivalent to the atmospheric pressures on the Earth at altitudes of above 37,500 meters (125,000 feet). At 13,500 meters on the Earth there is no longer enough pure oxygen to sustain human life, and at 18,000 meters blood boils at body temperature. Any human explorer on Mars will therefore require a full pressure suit. In fact, except for the Martian gravity, an astronaut would find a walk on Mars not much different from a walk in outer space.

The dust storms might create another hazard to astronauts. Martian dust is a fine grit, probably abrasive. On the Earth winds of 24 kilometers per hour can raise dust. The corresponding minimum wind velocity on Mars is about 160 kilometers per hour. Some Martian winds reach velocities of 480 kilometers per hour. The resulting scouring and sandblasting effects could be serious, although because of the low atmospheric pressure the astronauts would be in little danger of being blown away.

Some 60 percent of the dust observed by *Mariner 9* was silicon dioxide (SiO_2), or silica. Silicon compounds are thus obviously common on Mars. A number of the Martian volcanic formations are sufficiently similar to ones on the Earth to allow us to infer that they are composed of much the same materials, such as basaltic rock.

The red color of the Martian soil seen from telescopes on the Earth was one of the baffling questions that *Mariner 9* did not solve. The general opinion is that some kind of iron compounds are involved, but even the basalts on Mars were observed to have some red color as well—and terrestrial basalts contain no iron. It is certain that the old theory that the red color was due to large quantities of

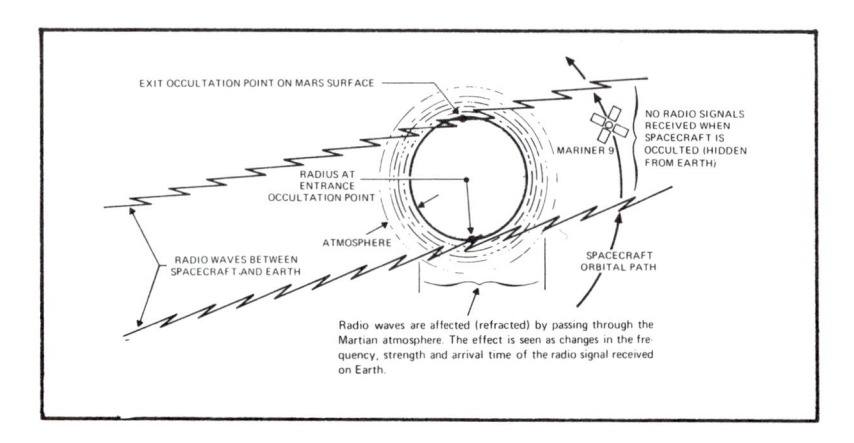

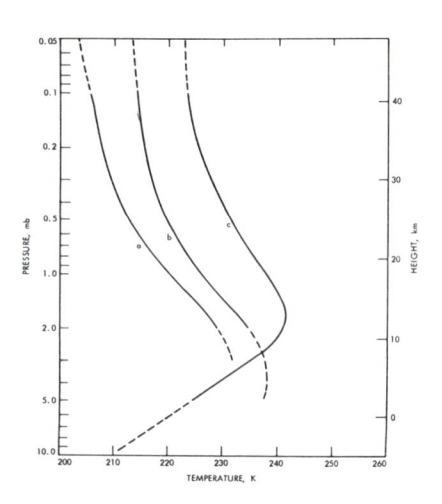

Temperature inversions on Mars. The curve on the right, labelled c, is from the south polar region, and shows a definite temperature inversion. That is, the temperature near the frozen icecap is lower than the temperatures at altitudes up to ten kilometers above the ice.

Measurements of the Martian atmosphere's temperature, pressure and density were made by passing Mariner 9's *radio signals through the atmosphere while the spacecraft was moving behind the planet. The atmosphere's refraction of the radio signals allowed scientists on Earth to make precise measurements of the Martian "air."*

iron oxide on the Martian surface is not true, but no good alternate explanation has yet been found. The sources of the colors in the lunar soil were not suspected until the *Apollo* missions brought back samples: no one had guessed that the lunar colors were due to the presence of titanium glass. An explanation of the Martian coloring will probably have to wait until samples are analyzed.

Because *Mariner 9* did not go into polar orbit we have incomplete information about the Martian ice caps. However, we are now certain that the part of the caps that disappears in the summer and reforms in winter is made of nearly pure carbon dioxide. The caps are thin, probably less than a few meters thick on the average. *Mariner 9* arrived in late spring of the southern hemisphere, a nearly ideal time for watching the southern ice cap vanish. Unexpectedly, the outline of the ice cap remained throughout the summer, long after any carbon dioxide should have sublimated. That fact suggests to Murray that underneath the cap of dry ice there is a residual cap of water ice.

The land surrounding the ice caps of both poles showed curious features that Murray calls laminated terrain: very thin layers of soil, alternately light and dark, looking much like "fallen stacks of poker chips." Groups of 20 or 30 of the laminas cluster together to make up plates some half a kilometer thick and as much as 200 kilometers across.

Murray and his star graduate student at Cal Tech, Michael C. Malin, are convinced that the series of overlapping plates were formed somehow during the cycles of condensation and evaporation of the polar caps. Curiously enough, the plates do not surround the poles in concentric circles, leading Murray and Malin to speculate that the arrangement of the plates was caused by changes in the tilt of Mars's axis of rotation.

According to Murray, the direction of Mars's axis changes owing to inbalances caused by internal convection currents within the Martian core, and the plates form around each successive position of the poles. Murray further believes that the eccentricity of Mars's orbit changes over geological time scales, causing the amount of sunlight received on Mars to vary and thus affecting how much of the ice cap is melted and reformed and how thick the plates become. The laminations and the areas of light and dark are produced by dust deposited during wind storms.

If the theory of Murray and Malin is correct, the polar regions are a record of how much sunlight Mars has received over the past 100 million years. Their theory is not universally accepted; the alternate view is that the lamina were caused by some process of erosion rather than formation.

THE GEOPHYSICAL HISTORY OF MARS

The primary technique for dating Martian features is crater counting; the density of craters is very important for dating Martian geological events. Present theory is that the period during which Mars was heavily bombarded ended several billion years ago, and there have been much less frequent impacts since.

Ancient craters are common over the southern hemisphere, but fresh young craters are rare. Thus the high-standing heavily cratered rocks in the middle and southern Martian latitudes are the most ancient features on the planet. Next come the large lava basins, which could only have been formed after Mars's interior had heated enough to give rise to a molten core. Hellas is the oldest of the Martian maria, and its rim has been almost completely destroyed by meteorite impacts. The other basins were formed later, and after this time the bombardment continued at a decreasing rate.

Volcanic activity started about the time that the basins came into being. The oldest volcanic feature of Mars seems to be a heavily impacted shield volcano on the northeast rim of Hellas. Next, plains areas such as those around Olympus Mons were formed along with the volcanism of Olympus Mons itself. Other plains were formed later as were at least some of the older closed-end troughs that have cratered floors. The upper part of Olympus Mons and the shield volcanoes on the Tharsis ridge probably formed next, contemporary with other younger troughs.

This chronology is generally accepted, but like every other theory about Martian geology it has its problems. Recent *Mariner 10* observations of the planet Mercury indicate that cratered terrain on Mars is far older than we supposed. Because Mars is in the asteroid belt it was first believed that the impact rate would be higher than for Earth, the Moon, and Mercury. This is now believed to be untrue. The evidence supports the view that Mars was bombarded at a rate always similar to that of the inner planets.

THE MOONS OF MARS

Mars has two satellites. Both appear to be similar to asteroids and may have been captured some three to four billion years ago. They are the darkest objects ever photographed in the solar system, probably darker than anything you have in the room with you at the moment. Their orbits are nearly circular, and preliminary results suggest strongly that they are tidally "locked" onto Mars as the Moon is to the Earth, rotating on their own axes once with each revolution around Mars.

The farthest out is Deimos at a distance of 23,500 kilometers. Its period around the planet is 30 hours 18 minutes. Deimos is quite small, only 16 kilometers long by 11 kilometers wide. Seen from the surface of Mars, it would be a point of light, not a disk, and would appear to be about as bright as Mars or Venus seems to us.

Phobos is considerably closer to Mars, only 9,300 kilometers away. It, too, is irregularly shaped, and is 27.2 kilometers long by 19.4 kilometers wide. It has a period of

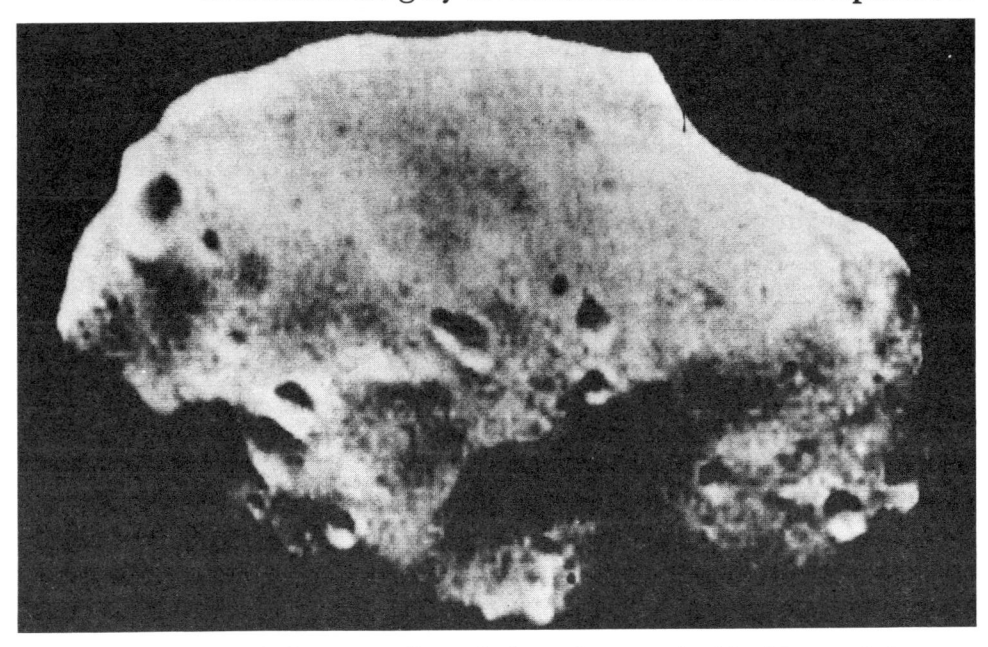

Mars's larger satellite, Phobos, photographed by Mariner 9 *from a distance of 5540 kilometers (3444 miles). Phobos is less than fifteen kilometers long, while Mars's smaller satellite, Deimos, is about half that size. Phobos appears to be little more than a large rock, although the craters suggest that it possesses considerable structural strength.*

7 hours 40 minutes, and although like Deimos it circles Mars in the direction of Mars's rotation, it travels so fast that it would appear to rise in the west. Near the horizon it would appear to be about 1/60 as bright as our Moon; directly overhead it would be closer, and it would thus appear to be 1/20 as bright as our Moon and about a third its diameter. It would go from the western horizon to the eastern horizon in about 5-1/2 hours with phase changes visible as it moved.

Both Martian satellites are heavily cratered by repeated meteroid impacts. Because there has been no erosion possible, the cratered moons have been useful in estimating the rates of impact on Mars itself and thus the age of cratered terrain.

IS THERE LIFE ON MARS?

Before the Viking landings, there was not one single observation indicating that there might be life on Mars. For all the data we had there could as easily be life on the Moon. Therefore it was meaningless to ask if life does exist on Mars. All we could ask was: are there now or have there ever been conditions on Mars that could be favorable to life?

Mars is cold. Its surface temperatures range from somewhat above freezing on summer days to below the freezing point of carbon dioxide (-125° C) on the polar ice caps. Equatorial midsummer afternoons might be quite balmy, reaching temperatures of some 30° C (80° F). The day-night variations are extreme, however, and even in the midsummer equatorial regions the nighttime temperatures will be 100° F below zero.

Moreover, the Martian surface is ceaselessly bombarded by ultra-violet radiation from the sun. During solar storms there is little atmospheric protection. Any life that could exist on Mars must be hardy indeed.

The atmospheric density is so low that any complex plant forms would need methods of processing large volumes of carbon dioxide: they must be an organic equivalent of a vacuum cleaner. Otherwise they could not take in sufficient gas molecules to work with. Except at equatorial summer noon the temperatures are too low for highly active water chemistry; the plants would need some form of internal heating, requiring more energy and also requiring that they process more carbon dioxide.

None of this rules out life as such, but life as we know it is extremely improbable under the present conditions on Mars. Thus speculation about the possibility of life on the Red Planet is inseparable from the problem of the geological history of the planet.

Shortly after the Martian "river bed" channels were observed, there was widespread speculation that Mars periodically goes through alternating "wet" and "dry" epochs, and that we are unfortunately observing Mars during the midst of some thousands of years of drought. Under this hypothesis, life may have formed during a time when there was denser atmosphere and open water, and then have adapted to increasingly hostile conditions. Such life forms may lie dormant under the dirt and grit of the channels, there awaiting the Great Thaw that will cover Mars with a reasonable atmosphere and bring water coursing through the dry river beds.

This theory is widely associated with Carl Sagan and the Cornell University group. In his *The Cosmic Connection,* (Doubleday Anchor, 1973; highly recommended) Sagan points out that due to the precession of the Martian polar axis and the eccentricity of Mars' highly elliptical orbit, the planet may experience dramatic changes of climate over relatively short time periods. He says, "twelve thousand years ago may have been the epoch of precessional spring and summer. The dense atmosphere of that time is now locked away in the polar caps. Twelve thousand years ago may have been a time on Mars of balmy temperatures, soft nights, and the trickle of liquid water down innumerable streams and rivulets..."

It is an attractive theory and was eagerly seized upon by both science and science fiction writers. Like Lowell's Mars, we *want* to believe in it.

Murray does not agree with this theory and says: "I always get to play the role of the heavy, shooting down attractive and popular theories." (Private interview, Nov., 1974.) "What Sagan says is wrong. It is easy to prove there have been no wet periods on any time scale of 10,000, or 100,000, or even 500,000 years. It is not so easy, however, to rule it out on a two- or three-billion-year time scale.

"The problem is there are features coexistent with the time the channels formed, and they have not been eroded. Therefore, it was not raining. You can't have large craters standing untouched while grand canyons are carved by flowing rivers."

Sagan replies: "Adjacent eroded and uneroded features seem to me to have no bearing on the question if the erosion is not uniform over the planet." (Private letter, November 1974).

One hypothesis of how the water is released during the presumed wet epochs is that local volcanism heats up ice locked under the Martian surface, releasing it somewhat in the same way that melting glaciers on Earth can produce enough water to carve extensive river beds. The difficulty is that the Martian pressure is so low that liquid water would quickly evaporate. Therefore a thicker atmosphere would be needed, and no one knows where that atmosphere could be stored during the dry epochs. It is not impossible to think of ways in which there could have been an atmosphere at one time; but how can it periodically appear and disappear?

It is not stored in the polar caps. They are simply not large enough. There could be large quantities of water locked in permafrost all over the Martian surface, but again it's hard to explain how the frost would melt only in some areas, do its work, and vanish without affecting nearby ancient features.

Malin gives a telling example in his doctoral dissertation (in preparation; private communication, Jan., 1975):

Emi Koussi is a volcano in the North African nation of Chad. It is a typical shield volcano, with channels formed by collapsed lava tubes, lipped craters, and so forth, and "looks very Mars-like." There are two kinds of features on Emi Koussi: those formed from lava flows and tectonic activities, and structures formed by water erosion. Since Chad is at present a desert area these latter must have been formed during the Earth's Pleistoscene Period when North Africa was a wet area.

The ancient features that formed in the early days of Emi Koussi's existence are well preserved—except where they are cut through by water erosion. Furthermore, it is known that this water erosion was relatively recent and took place over a geologically short time, that is, in the Pleistoscene.

Elysium Mons on Mars is larger than Emi Koussi, but it is very similar to its terrestrial counterpart. It is near the Tharsis shield volcanoes and is not greatly distant from some of the "riverbed" channels. It exhibits all the various features we see on the Earth: collapsed lava tubes, faulting and other tectonic effects, as well as some impact cratering. It shows absolutely no signs of water erosion. There is no gullying.

Malin and Murray see only three possible explanations of these features. First, Elysium Mons may be composed of radically different materials from Emi Koussi. Given the land forms around the two volcanoes, and the data obtained from the IRIS experiment, that seems extremely unlikely.

Second, the present surface features of Elysium Mons may have formed after the erosion events on Mars: the gullies are there, but they've been covered up. This also seems unlikely: all evidence indicates the surface is one or two billion years old.

Third, then, perhaps there are no erosion features on Mars because the cyclical periods of Earth-like atmosphere have never happened.

The Cal Tech group believes the "riverbed" channels may have been formed by flowing fluids—and water is certainly the most likely of these—but that they are very old, formed billions of years ago. They conclude that although it's impossible to rule out cyclical Martian "wet" periods, we can be positive there have been none in the past 100,000 years.

It is difficult to believe in life forms able to lie dormant for more than 100,000 years.

Bruce Murray is certain of only one thing: "we do not understand the geological history of Mars," nor do we have any good theories on how the channels were actually formed. "They may have nothing to do with water. There may be something unique to Mars, something bizarre. Certainly it didn't *rain*.

"There are huge volumes of material that we can't account for. Perhaps it subsided and was later extruded to form the Tharsis ridge—which is younger than most of the terrain around it. We just don't know.

"This isn't just disagreement among competing theories. There are no theories, none at all, that can explain all the existing facts. Something important has happened here that we just don't understand."

Whatever the explanation of the Martian channels is, it seems very unlikely that life evolved on Mars in any way similar to the way it did on the Earth. There seems to have been no long period in which Mars enjoyed liquid water and a thick atmosphere, both of which are required for life as we know it. Even if such life forms did evolve it is extremely unlikely that they are now lying dormant, waiting for more favorable conditions.

If we want to postulate Martian life, it will have to be life that can survive present conditions. For example, the life forms might be surrounded with silicon shells, creating their own environments with higher pressures and humidities. Glass is a good shield from some of the harmful effects of ultraviolet light, and of course it would raise internal temperature through a greenhouse effect.

Sagan says: "I wouldn't bet too much money against finding complex life forms there. There is no reason to believe or disbelieve that there are large organisms on Mars. Mars has had 4-1/2 billion years for independent evolution. Therefore the Martian organisms, if any, are not like us. We won't know until we land."

Others emphasize that there is no reason to believe that there is any form of Martian life at all. Murray goes further. "The search for life on Mars is another remnant of Lowell's Legacy. We don't think of looking for life on Venus or Mercury."

THE FUTURE OF MARS

Mars is a hostile place, but far less so than the Moon. Given transportation, Mars could easily be colonized; in fact, the logic of the situation would almost force long-term bases on Mars, since it is far easier to live there than to get there.

Given permanent scientific bases it would be no great effort to establish full self-sustaining colonies; and once those exist the planet will probably be transformed into a relatively pleasant home for several million people.

Terraforming Mars into a "shirtsleeves" environment is a task for generations. The goal is sufficiently long-term that it will probably appeal only to people already living there. On the other hand, colonization is simple enough that it is probably not unreasonable to expect a small number of scientific colonies within the next fifty years.

Why Mars?

Of the planets, Venus may be the easiest to terraform, and some experts believe most of the work could be done from Earth. Mars, however, is the most likely prospect for early colonization. The long-term prospects for Venus are probably better, but we cannot immediately land on Venus. The Moon is closer than either, of course, but establishing self-sustaining colonies there will be difficult, and lunar bases will probably always be dependent on Earth.

Mars, however, has everything needed for life. It receives enough sunlight to provide plenty of energy for colonists: in fact, the solar energy that actually reaches the Martian surface is not much less than what penetrates the Earth's atmosphere. Certainly there is enough to grow crops as well as to activate solar cells. The materials to make the solar cells also exist on Mars.

Mars has plenty of water; and given water and energy there is no problem with oxygen. Mars also has recent volcanism, indicating that metals and other building materials will be relatively accessible.

For all these reasons Mars is likely to be the site of the first completely independent colonies in the Solar System.

The initial bases on Mars will probably be little more than scientific laboratories. As stated earlier, it is senseless, given the costs of transportation, to plan "touchdown" missions to Mars; the first humans to land there should probably plan to stay a year or more, and to leave their bases intact for the next arrivals.

Early colonies on Mars will resemble those on the Moon, in that they will be underground, and humans going onto the Martian surface will need pressure suits. The suits are likely to be different from the present cumbersome Extra-Vehicular-Activity (EVA) garments. Recent laboratory developments show that a better suit would be a strongly elastic skin-tight garment something like a diver's wet-suit. The suit holds no pressure, and the traveller's skin reinforced by the elastic is the actual pressure wall.

Oxygen is supplied through either a pressure mask or a helmet attached to a neck seal. Temperature regulation is by the usual means on Earth: sweat. This requires no special equipment and is more reliable than most mechanical devices.

Large underground and domed colonies could be planted on Mars. Vehicles with pressurized cabins could be constructed from local materials once the first foundries and mines were set up. Except for being unable to go "outside" without special preparation, colonists would find life on Mars not very different from city life in extreme winter times.

The low gravity of Mars will probably prolong life: some space medicine authorities estimate that the normal life span will be well over 100 years. Colonists would also probably live at no more than half an Earth atmosphere; the lower pressure simplifies construction problems.

These colonies could become reasonably independent of Earth, since nearly every raw material needed already exists on Mars. Moreover, Martian bases could become the "capital" of extensive mining operations in the asteroid belt.

Asteroid mining has the great advantage that the interior portions of differentiated planetoids are easily accessible without deep mining pits. Pollution would hardly be a problem on a lifeless rock far from any center of inhabitation; while solar power for mining could be provided by large aluminized-mylar mirrors erected into paraboloid shapes. The absence of gravity makes such construction relatively easy.

Although asteroid mining depends on great improvements over present-day interplanetary transport systems, colonization of Mars does not; we could send long-term expeditions to Mars and keep them supplied with presently operational systems.

The first goal for Martian colonists would be to increase Mars' atmospheric pressure to about 1/4 Earth's, roughly equivalent to Earth at 35,000 feet. At that altitude it is possible to go outside the colony domes without pressure suits. Oxygen tanks would be required, of course, but it is easier to wear a mask than to carry the whole panoply of EVA equipment needed to survive present Martian pressures.

Another immediate advantage would be that liquid water could then exist on Mars. Existing water losses would be halted. Mars would begin to heat up, and the atmospheric temperature become more endurable. Plants could be grown without special environmental protection systems.

There are several ways to increase the Martian atmospheric pressure. All are highly speculative in that we do not know how much solid carbon dioxide is trapped at the Martian poles. However, the first step would be to melt the polar ice, either by direct means such as thermonuclear devices, or by coating them with dark materials so that they would absorb more heat from the Sun.

Since the amount of trapped carbon dioxide is unknown, the resulting pressures cannot be predicted, but there is probably not enough dry ice at the poles to bring the pressure to 35,000 foot equivalent. However, permanent melting of the polar caps will raise the pressures somewhat; that brings on a rise in temperature; and this makes available more heat for melting out other pockets of trapped dry ice.

Without further exploration it is impossible to know where the rest of the needed atmosphere will be found; and there is no agreement among planetologists as to how much may be available.

However, it is not unreasonable to suppose that between trapped polar ice, ice stored in the permafrost over the whole surface of the planet, and gases in the Martian interior, there will be enough carbon dioxide to bring the pressure at the Martian surface to between 35,000 and 10,000 foot Earth equivalent.

This will largely be carbon dioxide. The next step will be the introduction of plant life suitable for transforming the carbon dioxide into oxygen. Such plants are available, of course, and were largely responsible for Earth's oxygen atmosphere.

The colonies should have long-term prospects in mind. By long-term I mean hundreds of years into the future, when mankind will presumably either have exterminated himself on the Earth or will have available plenty of energy for large-scale tasks.

If the gases trapped on Mars prove insufficient, an alternate means is available. It is known that methane and water ice, as well as solid carbon dioxide, exist in large quantities in the outer Solar System. Given energy for propulsion there is no reason why some of these ice balls cannot be brought into an orbit that impacts Mars.

After the rather spectacular meteor display accompanying the impact ends, the imported gases could more than compensate for whatever shortages there were to begin with.

Even after terraforming, Mars is unlikely ever to have extensive seas. It will always be a dry planet, and the atmosphere will probably remain thin.

There will be no canals, of course. But except for the canals, the long-term prospects for Mars were described adequately in Robert A. Heinlein's *Red Planet* back in 1949:

"Climate—Similar to Alta Himalaya, extremely cold, extremely dry, air very rarefied. Temperature sometimes above freezing in daytime. Suitable for rugged life forms such as lichens and men..."

VIKING'S SEARCH FOR LIFE ON MARS

The *Viking 1* lander touched down on the surface of Mars at 11:53 A.M. Greenwich Mean Time, 20 July 1976—seven years to the day after the first men set foot on the moon.

On the rock-strewn sandy plain of Chryse Planetia the local time was 4 P.M., seven days past the beginning of summer. The distance between Earth and Mars at that moment was 341.5 million kilometers; radio signals travelling at the speed of light (300,000 kilometers per second) took 19 minutes to reach the jubilant scientists at NASA's Jet Propulsion Laboratory in California.

Six and a half weeks later, on 4 September 1976, the *Viking 2* lander touched down on Utopia Planetia, some 7500 kilometers east-northeast of the *Viking 1* lander. *Viking 2* found surface conditions almost identical to those at the Chryse site, although the rocks scattered across the plains of Utopia appeared to be more of a volcanic origin than those at Chryse.

The most important new discovery of *Viking 2*, however, was made by that vehicle's orbiting spacecraft, which determined that Mars's North Pole consists entirely of water ice, rather than frozen carbon dioxide. This means that water is abundant on Mars, even though it does not presently exist on the surface in liquid form.

The description of surface conditions on Mars given by science-fiction writer Robert A. Heinlein in 1949 turned out to be uncannily accurate. *Viking*'s instruments found that early summer on Mars is indeed "extremely cold"—temperatures varied from overnight lows of -85°C to daytime highs of -29°C (-121° F to -21°F, respectively). The air is "very rarefied"—only about 0.7% of normal sea-level pressure. And it is "extremely dry"—there is about ten-thousand times less water vapor in the Martian atmosphere than in Earth's, and it has not rained anywhere on Mars for thousands, perhaps millions, of years.

Yet despite these rugged conditions, *Viking* has produced evidence that comes tantalizingly close to confirming that life does exist on Mars.

(right) Viking *lander's extensible arms scoops soil sample (left) and deposits it in biology experiment (right) during laboratory tests. (Martin Marietta Aerospace photo) (below left)* Viking 1 *photo shows rock-strewn plain of Chryse. Horizon is about three kilometers (1.8 miles) away. (below right) Arrow points to dark debris kicked up when* Viking 1's *protective cover was ejected from the spacecraft and hit the Martian soil.*

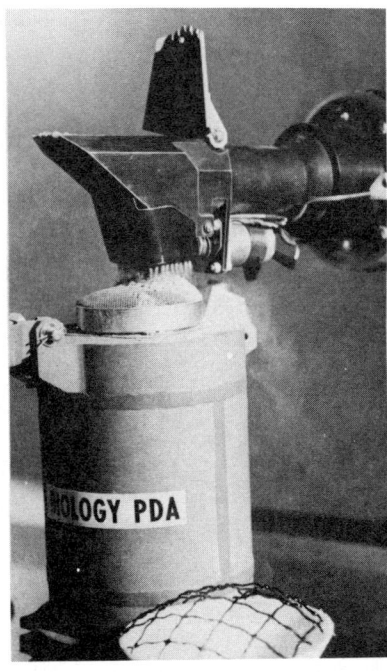

Stereoscopic cameras on Viking lander allow scientists on Earth to draw detailed maps of Martian surface. White lines trace irregularities of surface. (above) A closeup look at Mars' weathered, rocky surface. (below)

Panoramic view of ChrysePlanetia taken by Viking 1 *cameras shortly after touchdown.*

In fact, everything that *Viking* has told us tends to both confirm earlier theories of Mars and throw in several surprises.

To begin with, the *Viking* orbiters—the part of the complex spacecraft that remains in orbit around the Red Planet—photographed clouds of water vapor, and haze and fog lying close to the ground. This means that even though Mars is exceedingly dry by terrestrial standards, there is some water available in favorable locations for living organisms to utilize.

Instruments on the *Viking* landers sampled the atmosphere during the craft's descents, and then on the surface. Although the Martian atmosphere is 95% carbon dioxide, as expected, there are also unexpected traces of nitrogen and free oxygen, both vital to life as we know it here on Earth. Moreover, there is a surprisingly high amount of the inert gas argon—as much as 2%. Scientists such as Carl Sagan surmise that this argon is the remainder of a much thicker atmosphere that once girdled the planet, eons ago, when conditions may have been much warmer, wetter, and more hospitable for native Martian life forms.

Then came the pictures from the surface.

The incredibly clear, sharp photos of the Martian

surface proved that Mars is much more like the Earth than the barren, airless Moon. Sand dunes and rocks predominate the view. There is no vegetation in sight, no sign of movement that might be due to animal life. Yet Mars does not look that much different from many deserts on Earth. It appears somewhat similar to the most inhospitable stretches of Death Valley, if you can imagine Death Valley having the climate of Greenland or Antarctica.

And the Martian sky is pink, not the deep blue-black that had been expected. Red Martian dust, blown by the winds, tints the sky from horizon to horizon. There are clouds in that pink sky, but they are very few and thin. Every day is sunny on Mars.

It has been the biology experiments aboard the *Viking* landers, of course, that have provided the most excitement and the biggest puzzles.

The scoop at the end of each lander's extensible arm dropped tiny samples of Martian soil into each of three biology experiments:

1. *Label release*: Radioactive carbon 14 is introduced into the soil sample along with a nutrient mixture designed

U.S. flag is on housing of Viking's nuclear power system. The boxlike structure in the middle of the picture contains the seismometer. The view here is westward on Chryse Planetia. The hill in the center may be the rim of a crater. (left) *The view northeast from Viking 1 shows large boulders. The dark boulder in the center is about three meters (10 feet) across and three meters high.* (right)

Utopia Planetia, as photographed by Viking 2 *on the afternoon of 5 September 1976. Although also dominated by rocks, this area appears somewhat different from* Viking 1 *landing site.* (above) *High-resolution photograph from* Viking 2 *lander shows rocks at Utopia appear more porous and sponge-like than those at Chryse, suggesting a volcanic origin for them. Wind or liquid erosion has etched out channels in the soil between the rocks.* (below)

to "feed" any microorganisms that might be in the soil. If there are organisms present, the gases given off by the soil after a suitable incubation period should show a relatively high percentage of carbon 14.

2. *Pyrolitic release*: The soil sample is mixed with gases that simulate the Martian atmosphere (mostly carbon dioxide), again dosed with a slight amount of radioactive carbon 14 as a "tracer," and then exposed to a fluorescent light that simulates sunshine. If there are any Martian organisms in the soil that depend on photosynthesis, as terrestrial plants do, they should multiply vigorously in such an environment. After five days, the gases are flushed out of the chamber and the soil is baked at a temperature of 625°C (1157°F). If the baked soil releases a significant amount of carbon 14, it means that there were organisms in the soil that did take in the gaseous carbon dioxide.

3. *Gas exchange*: In this experiment, the soil sample is "incubated" in a water-rich nutrient bath and an atmosphere of carbon dioxide, krypton and hydrogen. Living organisms, if they are anything like life on Earth,

would convert some of these gases; for example, some of the carbon dioxide would most likely be converted to oxygen.

Each of these experiments has shown results that lie somewhere between what would be expected of lifeless, "straight" chemistry, and what would be expected of Earth-type biology.

For example, the gas exchange experiment showed a sudden increase in the amount of oxygen given off by the soil. But this soon levelled off, leaving puzzled scientists to wonder if the results were due to Martian biology or to some form of exotic chemistry that does not involve life at all.

Both the pyrolitic release and label release experiments showed similar behavior: startlingly strong returns, indicative of life processes, in the earliest results, followed by lower returns afterward.

Even the most skeptical scientists are now deeply curious about these extraordinary results. Clearly, there is something going on in those soil samples that is more than ordinary chemistry. Yet the results are not what one would expect from the kinds of biology we know here on Earth.

Perhaps this is exactly what should be expected of extraterrestrial life—neither "straight" chemistry nor

Mars, as seen by Viking 2 *on 5 August 1976, when the spacecraft was some 400,000 kilometers (250,000 miles) from Mars—about the same distance as the Moon is from Earth. The huge volcanic craters are in the Tharsis region, poking up above the morning haze and clouds. (left) "X" marks the spot for* Viking 2's *landing site, selected from photographs made by the orbiting spacecraft. The originally-selected landing site was rejected when closeup photos showed it to be too rocky for a safe landing. (right)*

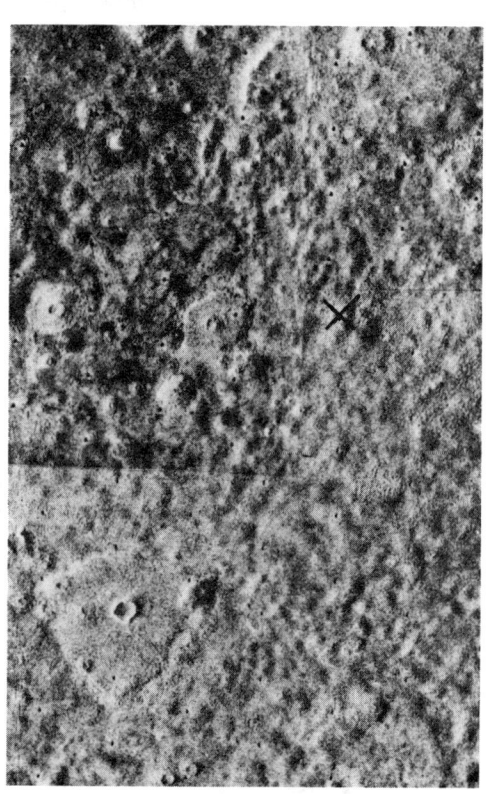

Earthly biology. Perhaps the experiments on *Viking* have indeed found native Martian life, but it's too different from our own forms of life for us to recognize from such a remote distance. No trace has been found, however, of organic chemicals in the soil of Mars. This is the most serious evidence against native Martian biology.

Carl Sagan sums it all up neatly: "For the first time nitrogen...[has] been detected in the Martian atmosphere. Nitrogen is the last remaining physical prerequisite for a Martian biology built on familiar terrestrial principles. Carbon dioxide is the major constituent of the atmosphere; water is available in vast quantities in the ices and minerals of the surface; and the planet is awash in sunlight."

In other words, the chances for life on Mars are very good—good enough for us to return to the planet with more sophisticated probes and, eventually, with teams of human scientists and explorers. *Viking* may not have given us definitive answers, but it has taught us which questions to ask the next time around.

But that is for the future. For now, science fiction writer Ray Bradbury gives the best explanation of the situation on Mars as of 20 July 1976:

"From this point on, there *is* life on Mars—an extension of our sensibilities. Man is reaching across space and touching Mars. *Our* life is on Mars now."

—Ben Bova
October 1976

ACKNOWLEDGMENTS   This article is based in large part on the MARINER MARS 1971 PROJECT FINAL REPORT (5 volumes; Jet Propulsion Laboratories Document, 1973-74) The author is grateful for generous donations of time by Dr. Bruce Murray and Michael Malin, both of the California Institute of Technology Division of Planetary Sciences. Thanks are also given to Dr. Carl Sagan of Cornell University for both interviews and unpublished data.

Conclusions in this article are the sole responsibility of the author. Interview quotations are given with the consent of those interviewed.

California Institute of Technology's Jet Propulsion Laboratories, Frank Colella, Director of Public Relations, have been immensely helpful in providing photographs, data, and reports.

Gregory Benford is Associate Professor of Physics at the University of California, Irvine. He is a theoretical physicist currently working in the areas of plasma turbulence, the dynamics of relativistic electron beams, and plasma astrophysics. His articles on science appear frequently in several national magazines, including *Natural History* and *Smithsonian,* and he is the author of a forthcoming text, *Life in the Universe.* He has published two novels (*Deeper Than the Darkness, Jupiter Project*) and his short stories have appeared in many anthologies. He lives in Laguna Beach, California.

The Exploration of Venus

Gregory Benford

VENUS	
mean radius	6056 kilometers
mass†	0.815
mean density	5.16 grams/cubic centimeter
surface gravity†	0.82
escape velocity	10.3 kilometers/second
length of day*	243 days 2 hours *retrograde*
length of year*	224.70 days
inclination of orbit to ecliptic	3°.4
inclination of equator to orbit	177°.8
mean distance from Sun	0.723 AU = 1.08 x 10^8 kilometers
eccentricity of orbit	0.007

satellites	diameter (km)	distance from planet (millions of kilometers)	orbital period (days)*	date of discovery
None	NA	NA	NA	NA

† Earth = 1
* Earth time

At sunset it beckons brightly on the far horizon, more brilliant than any star. The dot of light appears at the margins of the day, rising or setting more than three hours before or after the sun. In the early morning it is so bright the Greeks named it Phosphorus, meaning the light-bearer. They knew it as a planet, a wanderer in the sky. There are references in Greek literature to its alabaster

Venus, the Veiled Planet, as photographed from a distance of 720,000 kilometers (450,000 miles) on 6 February 1974, one day after Mariner 10's *closest approach to the planet. The spacecraft's cameras were equipped with ultraviolet filters which allowed details of the planet's cloud cover to be seen for the first time.*

light, compared to the white skin of a beautiful woman. Perhaps this is why Venus was always associated with the Goddess of Love.

History is full of ironies. Ancient shepherds watching the mottled pink and brown face of Mars connected it with blood and war. Yet Venus, seemingly serene and unchanging, is a far more hostile place. How we discovered this inversion of appearances and reality is a fascinating detective story, one that is still going on.

Men have squinted at the unblemished disk of Venus for centuries without learning very much. Galileo saw that the planet shows phases just as our moon does—this was a crucial point against the earth-centered theory of the solar system. It also explained why Venus is not much brighter when it is nearest the Earth, for then we see it only as a thin crescent. The planet rides about the Sun in a nearly perfectly circular orbit, unescorted by any moons. In the seventeenth century the Dutch physicist Huygens studied Venus for years and remarked "...she always appeared to me all-over equally lucid, that I can't say I observed so much as one Spot in her...is not all that Light we see reflected from An atmosphere surrounding Venus?"

It was an inviting speculation. Judging from its apparent size, the planet was slightly smaller than Earth. Its slight perturbing influence on the other planets led astronomers to conclude that it had about 0.81 of the Earth's mass. There was no way to tell how deep the atmosphere was, but reasoning by analogy with the Earth, astronomers estimated that at the surface the acceleration of gravity was about 88 percent of ours.

On the face of it, these numbers make Venus seem the most Earth-like of all the planets. Mars and Mercury are substantially smaller, and the outer gas giant planets are huge. Clouds seemed a very Earth-like feature; Mars has very few, and our planet is approximately half covered by clouds at any one moment. The fact that Venus was totally shrouded in a yellowish-white blanket might have indicated that there were some substantial differences between Earth and Venus, but the similar sizes and masses suggested the differences needn't be dramatic.

Our white billowy clouds are mostly water. It was natural to think that Venus was a wetter world than ours, cloaked in perpetual rainstorms. The Swedish chemist Arrhenius, a Nobel Prize winner, speculated in 1918 that "everything on Venus is dripping wet...a very great part of the surface of Venus is no doubt covered with swamps... constantly uniform climatic conditions which exist everywhere result in an entire absence of adaptation to changing exterior conditions. Only low forms of life are therefore represented, mostly, no doubt, belonging to the vegetable kingdom; and the organisms are nearly of the same kind all over the planet." With this recommendation by a world-famous scientist, the Venus-as-tropical-resort picture became popular. Some traces of the earlier Venus-as-Love-Goddess ideas remained, causing some to attribute the life forms beneath the clouds with benign, peaceful intentions. Without ever giving concrete reasons, popular articles assumed the intelligent inhabitants of our sister planet would never display the aggression of the Martians in H. G. Wells' *The War of the Worlds*. It was easy to assume the Venusians would copy our own best attributes, right down to the bicameral legislature and popular charities.

All this we speculated, without knowing how long the Venus day was, the surface temperature, or what the clouds were made of. In the 1920's several attempts to measure water vapor in the clouds failed to detect any. This led to the Venus-as-desert model, identifying the clouds with huge dust storms stirred up by high winds at the surface. The apparent absence of water freed speculation in other directions, too—some astronomers calculated that the surface might have thick oily oceans of hydrocarbons and that the clouds resembled smog. But in the absence of any further data, virtually any guess was as good as another. "It's embarrassing," a prominent astronomer commented in the late 1940s. "The closest planet to us—but we know less about it than we do about Jupiter, a world much further away. It may be like the Bahamas, or it could be sudden death if we stood on its surface. We just don't know. We can see it very clearly, but that white mask doesn't tell us a thing."

Whatever the clouds were, their lemon-yellow color meant that something in them absorbed blue light preferentially. Some 80 percent of the sunlight falling on Venus is reflected. For life-on-Venus advocates this was a hopeful result. Venus orbits about 40,000,000 kilometers closer to the sun than Earth, so it would be a hotter place unless it could get rid of most of the incoming sunlight. By reflecting 80 percent a simple calculation showed that Venus could have temperatures more or less like those on the surface of Earth. Of course, this is an estimate of the temperature at the top of the clouds, miles above the Venusian surface. The best calculated value (-47°C) agreed quite well with direct measurements of the infrared light from the clouds (-43°C).

This cool temperature for the upper clouds helped explain why earlier studies of the spectrum of Venus had found no water. Water vapor is easy to detect, but ice is not. Ice crystals in the cool cloud tops would escape observation, but carbon dioxide gas—which freezes only at a much lower temperature—would show up easily. The abundance of carbon dioxide seemed to argue against a Venus rich in life, because plant materials, animal bones and shells tend to absorb this gas from the atmosphere. But if water was present in the high ice clouds, it might also appear lower in the atmosphere. Where there is

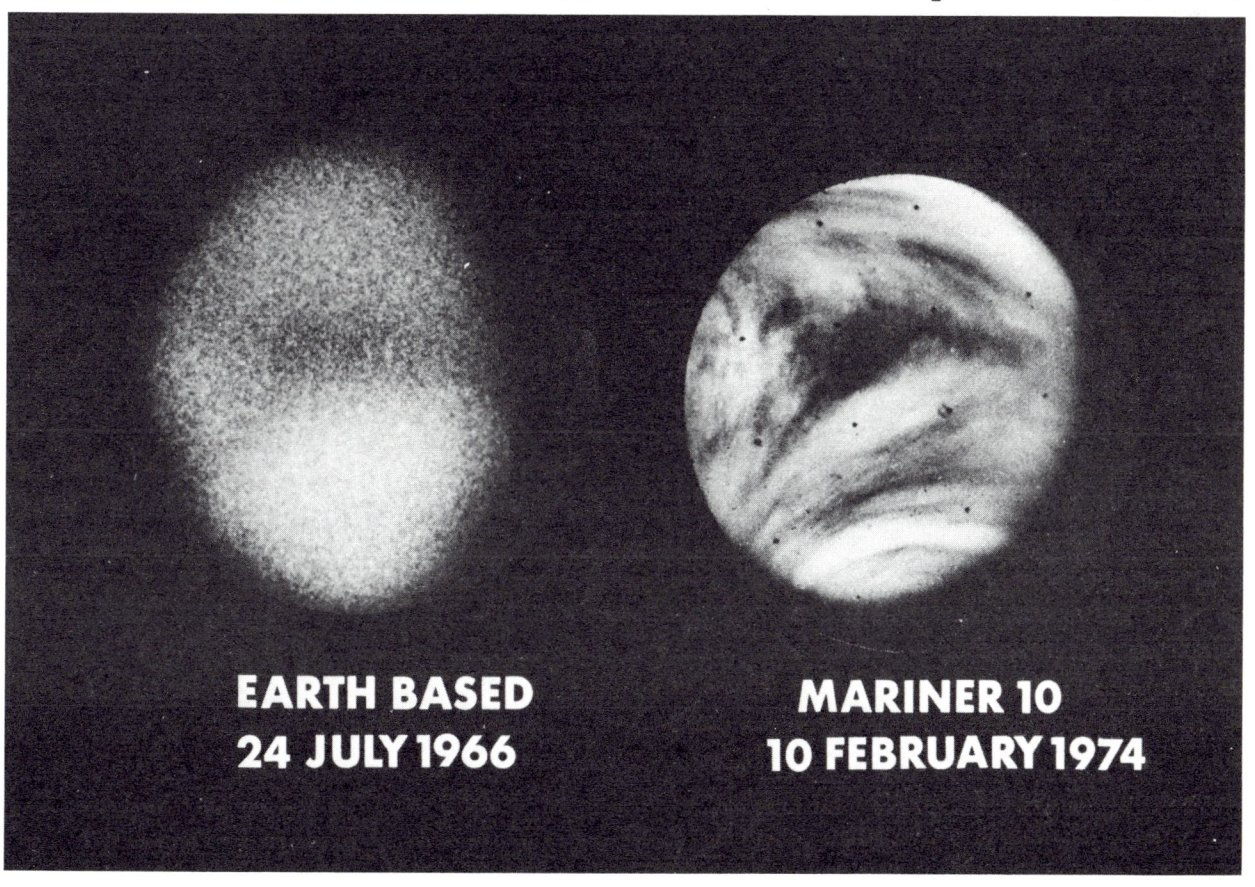

**EARTH BASED
24 JULY 1966**

**MARINER 10
10 FEBRUARY 1974**

Comparison of the best view of Venus obtainable by Earthbound telescopes with the view from Mariner 10. *Earth never gets closer to Venus than 41 million kilometers (about 25 million miles).*

water, there could well be life. The crucial remaining question was, how warm was Venus? The -43°C reading near the top of the clouds was reassuringly close to earthlike conditions, so things seemed hopeful.

All this changed in 1956. Every object gives off electromagnetic waves. The precise mixture of the wavelengths depends on how hot it is. If a body is very hot, some of its radiation appears in the visible spectrum; we use this in everyday life to judge how hot a flame might be.

Planets are warmed by the sun and in turn radiate away most of the energy in the infrared part of the spectrum. Many gases absorb infrared quite effectively, though, so the direct radiation from the surface of Venus was probably absorbed in the clouds. Whatever infrared radiation we measured from the planet probably indicated the conditions only near the top of the atmosphere. To look deeper into the atmosphere required studying a different wavelength of radiation. An obvious candidate was radio waves; the hotter a body is, the more energy it emits in the radio spectrum.

A group led by Cornell H. Mayer measured radio wavelength emission from Venus in 1956, using a parabolic antenna 50 feet in diameter. The waves they measured were three centimeters and 10 centimeters in wavelength. The astronomers waited until Venus was at its closest approach to the earth, in order to get the strongest signal from the planet. The amount of radiation received could be directly related to the temperature of whatever was emitting the waves.

Almost incredibly, the temperature indicated was 330°C, far higher than nearly everyone's expectations. What's more, there seemed to be little temperature difference between the dark side and the sunlit side of Venus.

There were further puzzles. Although the 330-degree temperature appeared in emissions at centimeter wavelengths, at shorter wavelengths in the millimeter range, there was a sharp decline in temperature to around 40°C. The easiest way to explain these differences was to assume that some of the radiation emitted by Venus was being absorbed in the planet's atmosphere. The simplest assumption was that the centimeter radiation came from the hot surface. If carbon dioxide was absorbing some of this radiation, the millimeter waves should originate at least 17 kilometers above the surface. The infrared region of the spectrum, which gave even lower temperatures, had to come from clouds nearly 35 kilometers above the hot surface. This was a very deep atmosphere. It was quite a forbidding environment for life as well, and several astronomers sought out alternate theories that could explain the data.

The crucial assumption of the hot-surface model of Venus for explaining the observations is that the centimeter radiation comes from the surface. Once that is admitted, the rest follows necessarily. Other models were possible, and they were pursued in the late 1950's. The question excited great interest because the quickly advancing space program promised further detailed measurements of Venus in the near future. Theories could be tested.

There are many things other than the surfaces of planets that emit great quantities of radio waves. Charged particles trapped in the Earth's magnetic fields do. Lightning flashes between clouds in our own atmosphere produce considerable radio interference. Charged particles in the dense upper atmosphere of Venus could, conceivably, radiate strongly in the radio spectrum. All these ideas were avidly pursued.

If the ionized layer in the upper part of the atmosphere of Venus (the ionosphere) was responsible for the centimeter radio waves, an alternate model followed naturally. A hot, dense ionosphere would be transparent to shorter millimeter waves. These millimeter waves could come directly from the surface, which would have to be between 50°C and 100°C. Then the infrared radiation would presumably come from cool clouds above the surface yet below the ionosphere. Thus Venus could be relatively mild at its surface, perhaps even cool enough for liquid water to form.

The hot-ionosphere model seemed reasonable, and in fact relied on the radio-emission properties that we knew worked in the Van Allen belts around the Earth and Jupiter. But in several details the hot-ionosphere model seemed shaky. Radar telescopes began probing Venus in the early 1960's and found several aspects of the hot-ionosphere model were not borne out. It was always possible, however, to get around these objections with a somewhat more complicated and ingenious model.

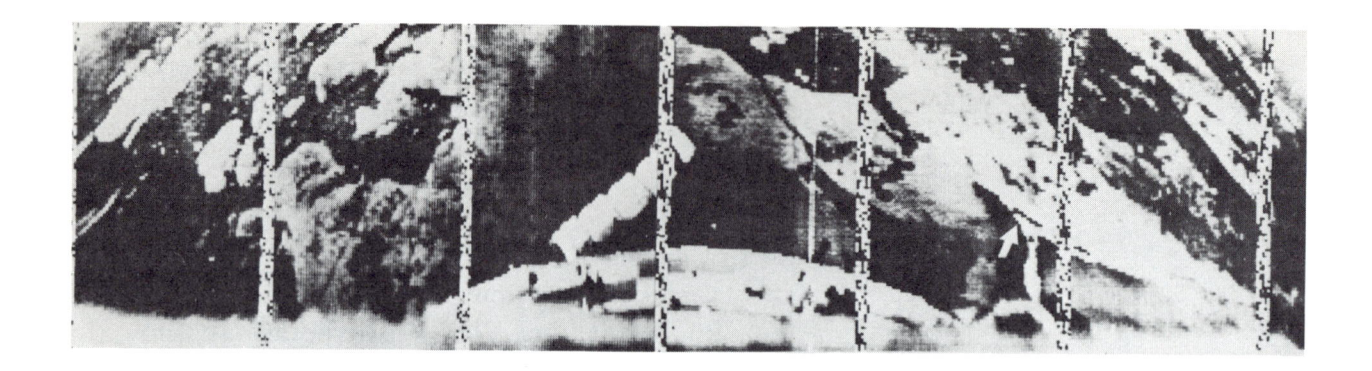

Nevertheless it seemed that the question could not be conclusively settled by a few simple measurements made from the Earth combined with elaborate theories. The stage was set in 1962 for a conclusive experiment that could only be carried out from a spacecraft passing near Venus.

The clearest experiment calculated to differentiate between the hot-surface and the hot-ionosphere models involved a small radio telescope. From a position close to Venus, the radio telescope could easily discern a difference between the emission from the center of the disk of Venus and emission from the edge. That difference would be important, because the hot-ionosphere model required that most of the radio emission (around one centimeter wavelength) should come from a layer of ionized atmosphere kilometers above the surface of Venus. When the radio telescope looks at the very edge of the planetary disk, it should register a greater thickness of ionosphere, with less emission from the surface below that layer. That means the radio telescope should register more radio emission at centimeter wavelengths from the *edge* of the disk than from the center.

The rocky surface of Venus, photographed by the Russian spacecraft Venera 9 (top) on 22 October 1975 and Venera 10, three days later. The "fish eye" effect is due to the camera's panoramic lens. Astronomers were surprised that so much light penetrated the perpetual cloud cover and reached the surface. (Photos by Tass, courtesy Educational Audio Visual, Inc.)

The picture would be just the opposite if the hot surface model were true. Centimeter-wavelength radiation is weakly absorbed by carbon dioxide gas, in which Venus' atmosphere is rich. Looking toward the edge of the planetary disk, the radio telescope would register the presence of more carbon dioxide absorbing the radiation than when it pointed toward the center of the disk. Thus there should be more radio emission from the *center* of the planet than from the edges. The result is called limb-darkening. If the hot-ionosphere model were correct, the edge (limb) of Venus should be brighter at radio wavelengths.

Mariner 2's microwave radio telescope showed Venus' limb to be "darker" in radio emission than the center of the planet, proving that the planet's surface was the source of the high temperatures.

Since the launching of the first *Sputnik* it had been obvious that the U.S.S.R. had much larger rocket booster systems than the United States. Therefore, they could launch immense vehicles of relatively little sophistication and great ruggedness. The United States had to rely on small satellites and planetary probes, expensively engineered to do the best possible job with the lightest, most durable electronics.

The competition between the U.S. and the U.S.S.R., each with different strengths and weaknesses, dominated the early days of the space program. The Soviet Union tried virtually every "launch window" that came with

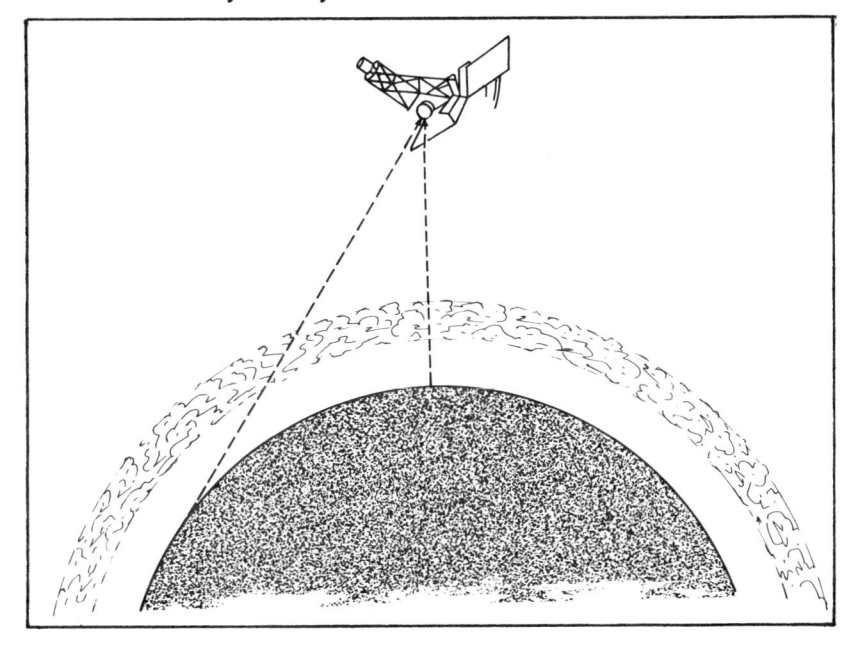

The Venus limb darkening experiment of Mariner 2. *As it flew past Venus, the spacecraft's sensors scanned the disk of Venus. Highest temperature came from center of disk, where atmosphere is thinnest, thus proving that Venus' high temperature comes from the planet's surface, not its atmosphere.*

each close approach to a nearby planet. The military interests in the Soviet Union apparently decided early to attempt the more spectacular coup. Consequently, by the end of 1972 five U.S.S.R. capsules had entered Venus' atmosphere. Similarly, there have been two Soviet landings on Mars.

The first attempt to send a manmade object to the vicinity of another planet was a probe, *Venera 1*, to Venus launched by the U.S.S.R. in February, 1961. In mid-flight the Soviets lost contact with it. They enlisted the help of the English Jodrell Bank radio telescope and tried to regain contact, but these attempts failed and *Venera 1* was never heard from again.

The United States followed soon after with the launch of *Mariner 2* on 27 August 1962. The launch was successful and the craft set out on its 293-million-kilometer arc toward Venus. It unfurled winglike solar panels to draw power. Its predecessor, *Mariner 1*, had gone off course during its launch from Cape Canaveral, Florida, and had

The Mariner 10 *spacecraft, developed at NASA's Jet Propulsion Laboratory and launched in 1973 to study Venus and Mercury.*

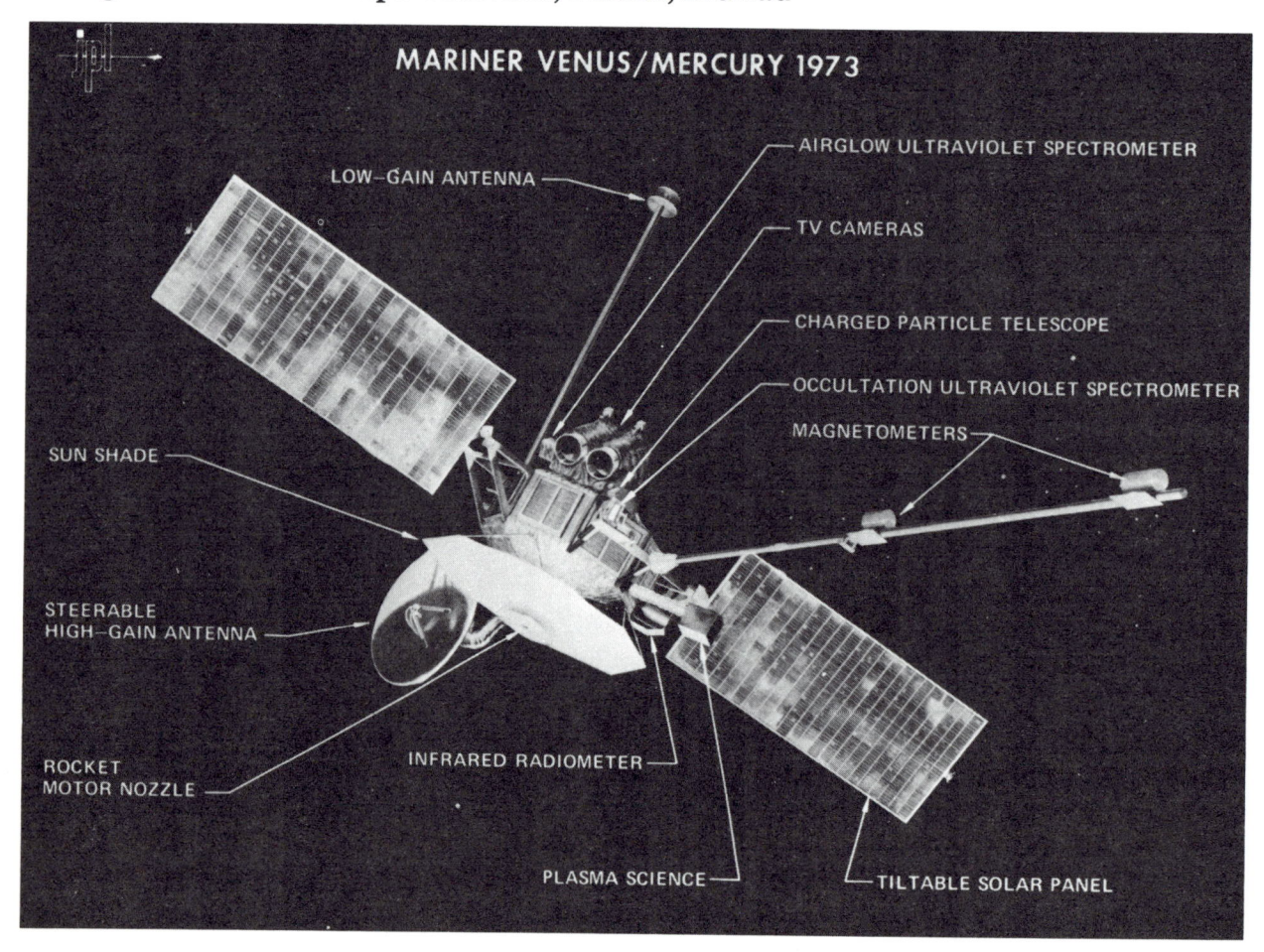

been deliberately destroyed. *Mariner 2* had to be assembled on only a ten-month schedule, meaning that the probe was a half-breed with earlier Ranger probes designed for exploration of the Moon.

Sending a probe from one orbiting planet to another is more complicated than simply firing it in a straight line. When an orbiting object slows down, the sun's gravitational pull draws it inward. So in order to leave the Earth's orbit and move inward toward Venus, *Mariner 2* first was fired opposite to the direction of the Earth's orbital speed. That reduced the speed of *Mariner 2* and caused it to slowly follow an ellipse inward toward Venus. Every planetary flight requires remarkable marksmanship. Launching a rocket from the surface of the Earth and then making mid-course corrections during the flight is a delicate task. It has been compared to sitting on a merry-go-round and shooting a bullet to hit a fast-flying sparrow beyond the horizon.

On its long flight *Mariner 2* made several measurements of the interplanetary medium, including a study of the solar wind, the fine spray of particles thrown out by the

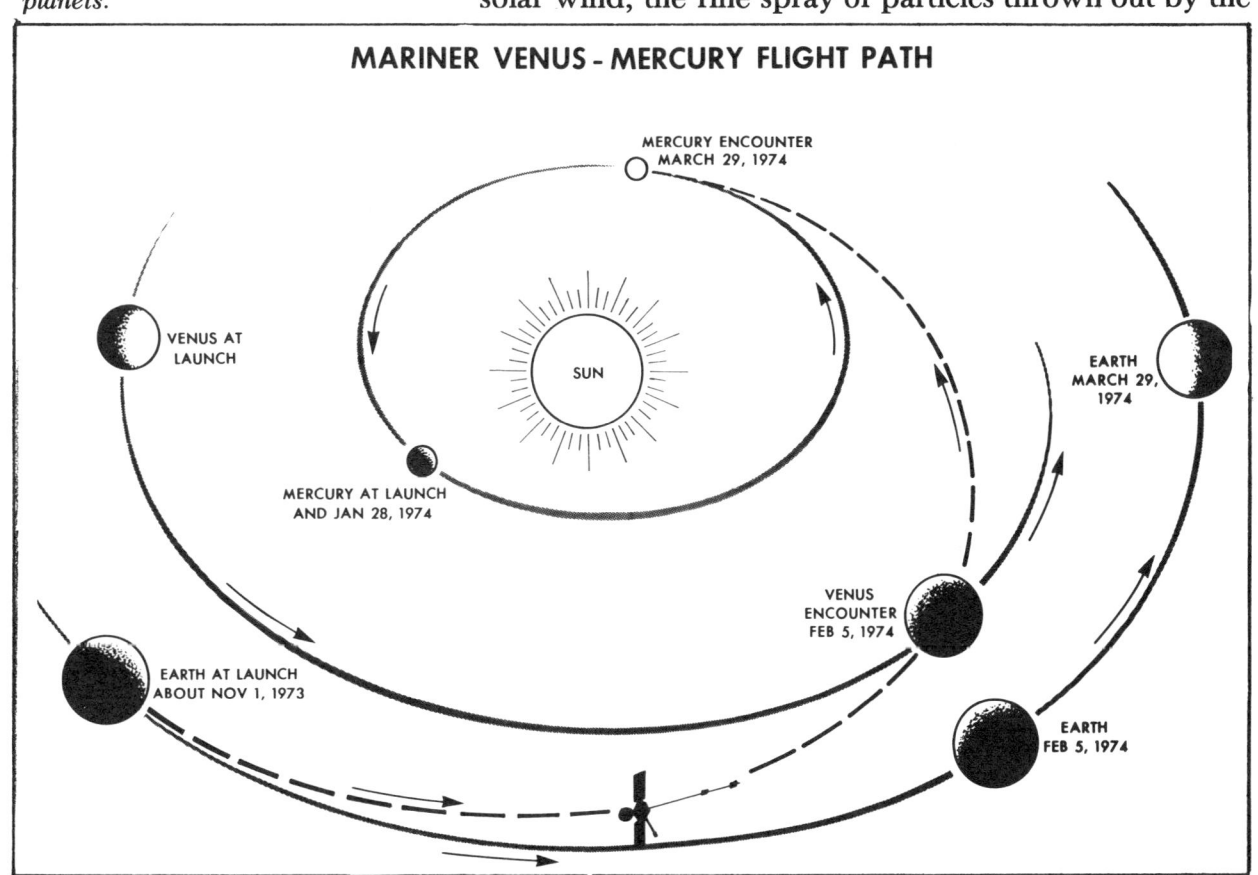

Mariner 10 *flight plan. This spacecraft was the first to explore two planets.*

MARINER VENUS - MERCURY FLIGHT PATH

sun. Part way through the flight something struck the craft with the speed of a bullet. *Mariner 2* shuddered, lost its sighting on the sun, and then regained it a few minutes later. What struck it was a meteorite weighing between 10 and 25 grams and moving at 500 meters per second. The craft was barely engineered to withstand such an impact, but it did.

As *Mariner 2* neared Venus, the sun seemed to swell a third again in size. Venus appeared to be a narrowing crescent and then, as the craft passed by, a dazzling white disk. The flight path of *Mariner 2* curved slightly, tugged by the mass of Shakespeare's "full star that ushers in the even."

For days before the fly-by, *Mariner 2* had been running a fever. The battery aboard ran as high as 55°C, far above the level at which it had been tested. At several points *Mariner 2* failed to turn on its instruments as it had been programmed to do. The probe activated itself again, however, under command from the 64-meter radio telescope in Goldstone, California.

Although the Earth was the third brightest object visible from *Mariner 2* as it neared Venus, the instruments on board ignored our blue-green dot. The radio telescope mounted above the center of the craft's spidery body nodded three times as the probe arced around Venus. It took a series of measurements on the night side, a sweep across the dawn line and a further scan across the dayside disk. The scans allowed a careful study of the planet's limb and a check to see whether or not the edge of the planet was brighter than the center at radio wavelengths.

When *Mariner 2* chattered its digital message back to the Jet Propulsion Laboratory in Pasadena, California, the hot-surface advocates carried the day. There was a distinct darkening at the edge of Venus' disk, indicating that radio waves came primarily directly from the surface. Venus was a cauldron whose temperature was somewhere between 130°C and 480°C.

Mariner 2 told us a good deal more about Venus in addition to settling the temperature dispute. It showed that Venus differs strongly from the Earth, which has belts of trapped energetic particles trapped in its magnetic field. Venus has less than a hundred-thousandth of the magnetic field of the Earth. The absence of any appreciable magnetic field was a mystery. We believe that the Earth's field comes from currents driven in the high-conducting metallic core. The Earth's center is squeezed by the matter above it, and heated by the decay of radioactive elements. Our 24-hour rotation period churns currents in this core and produces a magnetic field. Since Venus is only slightly smaller than Earth, it seemed in the early 1960's that it should have a magnetic field since it probably had a core. But even with a core, if the planet did not spin rapidly the electrical currents would be small. That indeed appears to be the case, for other lines of evidence have told us that the day on Venus is very long.

Even *Mariner 2* at close range could tell us nothing about the length of the day on Venus. All that we know about it comes from radar. When a pulse of radar strikes a planet, it is reflected in much the same way as a polished ball reflects light, so that if the surface of the planet is moving, the radiation returned to the radar antenna is altered. If a portion of the planet's surface is moving toward the radar antenna, the frequency of the pulse that returns will be slightly higher. If the part of the planet is moving away from the radar antenna, the reflected radar pulse returns with slightly lower frequency. If the planet is spinning, one side is moving away from the Earth and the other toward the Earth. Thus a slight up and down shift in the frequency of the radar signal can tell us how rapidly Venus is rotating. Because Venus hides behind its veil of atmosphere, this is the only way we can measure the length of its day.

Measurements from Earth in 1961 established that the rotation was slow, but no more than that. By the time *Mariner 2* was on its flight, the Goldstone facility had improved its sensitivity enough to find out in which direction Venus was rotating. Surprisingly, it is rotating backwards.

On the Earth, if we were to stand at the North Pole and watch the planet turn beneath us, objects would rotate in a counter-clockwise direction. The Earth revolves around the sun in the same manner, also counter-clockwise. Standing on the North Pole of Venus would give the opposite effect—surface features would appear to move clockwise. This is called retrograde rotation. Of all the planets in the solar system, only Venus and Uranus have this unexplained property.

Even more mystifying, Venus' day was shown by radar measurements to be 243.16 Earth days. This is so unlike both Earth and Mars that it raises questions immediately.

A clue may lie in the fact that this 243-Earth-day rotation period makes Venus present exactly the same face to Earth whenever it is closest to us. How Venus got this way is probably tied up with conditions at the origin of the solar system, and it may also involve a peculiar relationship between Venus and Earth. The fact that these two planets perform an elaborate waltz, with Venus always turning the same face toward us as it draws nearby, means that Earth probably has exerted considerable tidal influence on Venus, just as it does on our Moon.

Tidal forces operate effectively if there is a bulge in a planet. It is possible that Venus has a lump in its equatorial plane, which allows Earth to exert a considerable tug on it when the two planets are near each other. It has also been suggested that Earth's tides pull at Venus' heavy atmosphere. Over billions of years this interaction may have slowed Venus. There are other ideas—that Venus once collided with an asteroid about 200 kilometers in diameter, a catastrophic event that could also have reversed the planet's rotation. The problem remains open.

Radar measurements used in conjunction with optical observations led to a measurement of the depth of Venus' atmosphere. It is thick, between 44 to 65 kilometers deep. This heavy blanket of carbon dioxide means that although the planet's day is long, the night is not chilly. Instead, heated gas from the day side quickly blows around to the night side. This is why measurements of the night side of Venus have given temperatures just as high as on the day side.

Mariner 2 gave good indirect evidence that Venus was so incredibly hot at the surface that even lead would melt. But the evidence was indirect, and some felt it was not conclusive. This led the Soviet Union to make an attempt at an actual entry into Venus' atmosphere. Their *Venera 3* probe launched in 1965 did enter the atmosphere, but something failed in the probe and no data were returned.

In 1967 *Mariner 5* flew by Venus and the Soviet *Venera 4* dropped a capsule. The small, rugged brown ball deployed a parachute and sank slowly through the upper cloud decks. Part way down, the craft stopped transmitting. The Russian scientists interpreted this as an impact with the surface. But the reported temperature of about 280°C and the surface pressure of 20 times the pressure at sea level on the Earth disagreed with interpretations of the data from *Mariner 2*.

VENERA 8 LANDING CAPSULE

TASS drawing

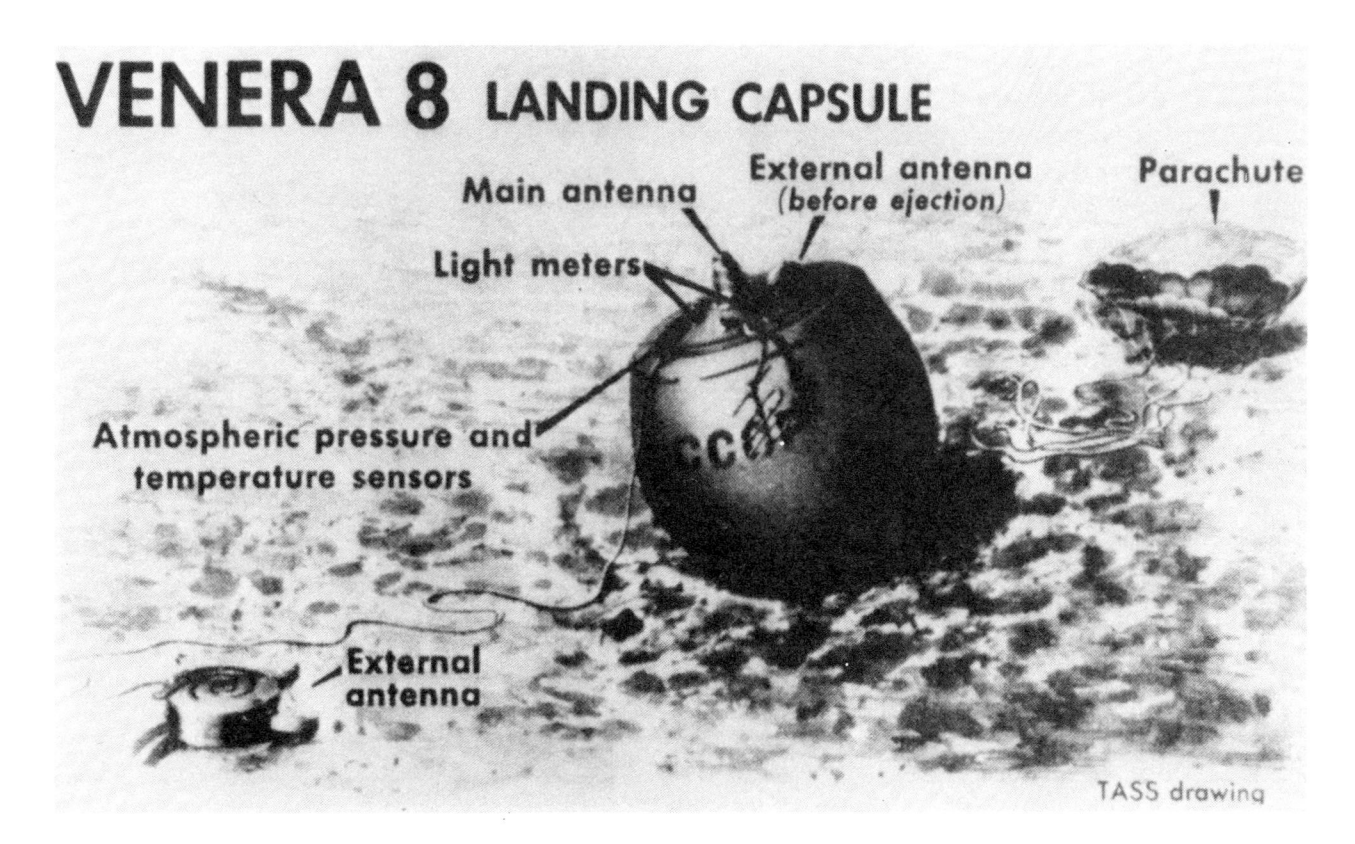

Artist's rendering of the Venera 8 *spacecraft on the surface of Venus. This Russian craft was the first to land successfully on another planet and transmit data on surface conditions back to Earth. (Tass drawing)*

This point came to a head at a 1968 meeting in Tokyo. Faced with data from both *Mariner 2* and *Mariner 5* and some convincing arguments framed by the Americans, the Soviets argued that *Venera 4* had landed atop a high mountain. This would have meant, though, that the craft happened to set down on a 24-kilometer-high mountain. At the same time, radar studies showed no mountain on Venus higher than a couple of kilometers. The Soviet explanation seemed too unlikely. Cornell astronomer Carl Sagan reports that a Russian professor replied to these comments by asking what Sagan thought the probability was that the first German bomb to fall on Leningrad in World War II would kill the sole elephant in the city. Sagan responded that the chance was small. The professor chuckled and pointed out that in fact that was precisely what happened. The elephant in the Leningrad zoo was the first casualty. So huge coincidences do occur.

Elephants aside, it seemed more likely that *Venera 4* had simply collapsed under enormous pressure. Despite the findings of *Mariner 2*'s limb-darkening experiments, the Soviet designers of *Venera 4* had believed the implications of the hot ionosphere model, which indicated the atmospheric pressure near the surface should not be terribly high.

93

The U.S.S.R. designed a later craft, *Venera 7*, to withstand pressures 180 times that at the surface of the Earth. In December of 1970 it parachuted to the parched, rocky surface and sent back radio signals for twenty minutes before it was fried. These signals indicated that the atmospheric pressure of Venus is 70 times that of the Earth. Temperatures reach 400 degrees Celsius. Brisk winds whip around the planet, carrying heat from the day side to the night.

Between 1965 and 1972 the Soviet Union attempted a total of ten landings on Venus. Four entered the atmosphere and two reached the surface. *Venera 8*, a special design, made several varied geological experiments. *Venera 8* was encased in a descent module that was dropped from an orbiter. The module fired braking rockets and was also slowed by atmospheric friction, sending the temperature of the shell to over 12,000°. Its descent by parachute took about an hour. A small port opened in the side of the round craft, admitting atmospheric gases in order to measure the concentration of ammonia. A fine-grained yellow powder was exposed to the atmosphere and turned blue; the color change was

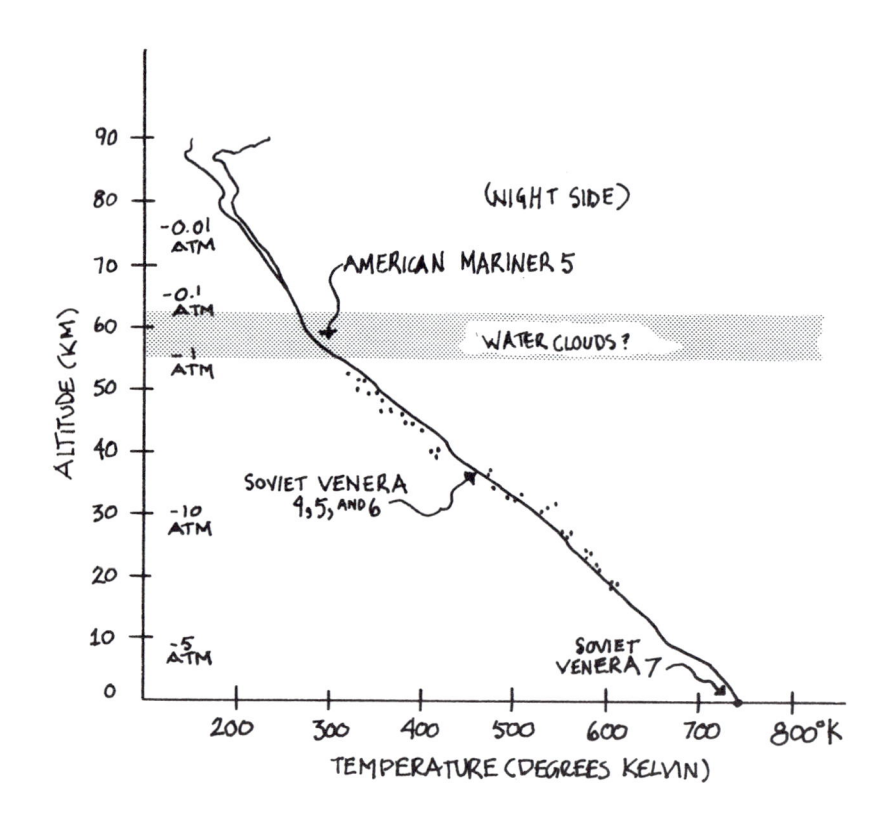

Temperature distribution in the lower atmosphere of Venus as a function of altitude.

registered by a simple light-sensitive circuit. The test showed that there was at least 0.01 percent ammonia in the lower reaches of Venus' atmosphere. Similar tests found the atmosphere was 97 percent carbon dioxide, not more than two percent nitrogen, less than 0.1 percent oxygen, and had less than one percent water vapor just beneath the cloud layer. *Venera 8* landed softly amid loose granite rock. It survived the intense 470°C temperature for 50 minutes. It was designed to do this— the components had been frozen while the craft was still in orbit, so that it could survive longer on the surface. *Venera 8* touched down on the sunward side on the planet, but before the sun had risen over the horizon. A small light-sensitive device registered a kind of dim twilight, showing that even Venus' thick cloud blanket does not filter out all sunlight.

The U.S.S.R.'s *Venera 9* and *Venera 10*, both of which landed in October 1975, outlasted all earlier Venus probes by lasting 53 and 65 minutes, respectively, on the surface. Some scientists had thought that the cloud-dimmed light would make pictures impossible, or the dense atmosphere would distort images too much. But the *Venera 9* photos, the first from the surface, showed angular rocks, unblunted by erosion of wind or sand. This immediately disputed the conventional wisdom that the thick atmosphere would cause even slow winds to grind down surface features.

Even more surprisingly, the sharp rocks cast sharp shadows. Unless the landing craft carried its own lighting—which the Soviets did not mention—the shadows must come from sunlight, even though the thick atmosphere should diffuse this light.

Venera 10 sent back pictures strikingly different. It saw a horizon 200 to 300 meters away, with rounded and flat rocks testifying to strong erosion. This difference from *Venera 9* may mean that Venus varies considerably from spot to spot. Sharp-edged rocks may mean that volcanic eruptions or quakes produced them. If so, Venus is internally active.

These first photos are tantalizing, raising more questions than they answer. They testify that Venus remains, at least in part, a mystery.

This is a bleak, tortured picture, but one not without interest. Even with an atmosphere 90 times denser than Earth's, Venus apparently cannot shield its surface from large meteorite impacts. At least, that is a preliminary conclusion from detailed radar studies done by Richard M. Goldstein and his collaborators at the Goldstone Deep Space Station in the Mojave Desert.

From the radar studies there is evidence that in the center of Venus there is a 1,470-kilometer circular swath, about the size of Alaska, pitted with a dozen huge, shallow craters. The largest is 160 kilometers across but only 400 or 500 meters deep.

"Admittedly," Dr. Goldstein says, "there is a certain symmetry to the pattern. The largest crater has a very close twin on the other side of the dark bar. This seeming mirror image of the crater gave us quite a bit of worry before we were through analyzing. But I believe it is there."

The size and shallow nature of the craters suggests that they are impact craters, not volcanic, though the question is by no means closed. To survive the fall to the surface through the thick atmosphere would require a very large meteorite indeed. How these craters could persist despite the rapid erosion by super-heated, dense, dust-ladened

(left) *Radar map of Venus produced by astronomers of the Jet Propulsion Laboratory. Since Venus is perpetually covered with heavy clouds, radar is the only way to "see" the surface... except for landing spacecraft. The circle shows a 1500-kilometer-wide area that contains numerous large craters.* (right) *Closeup view of the circled area from the previous picture. Craters up to 160 kilometers across are visible. The black area through the center of the scan is a belt that cannot be accurately tested from the Goldstone Tracking Station, where the radar measurements were made.*

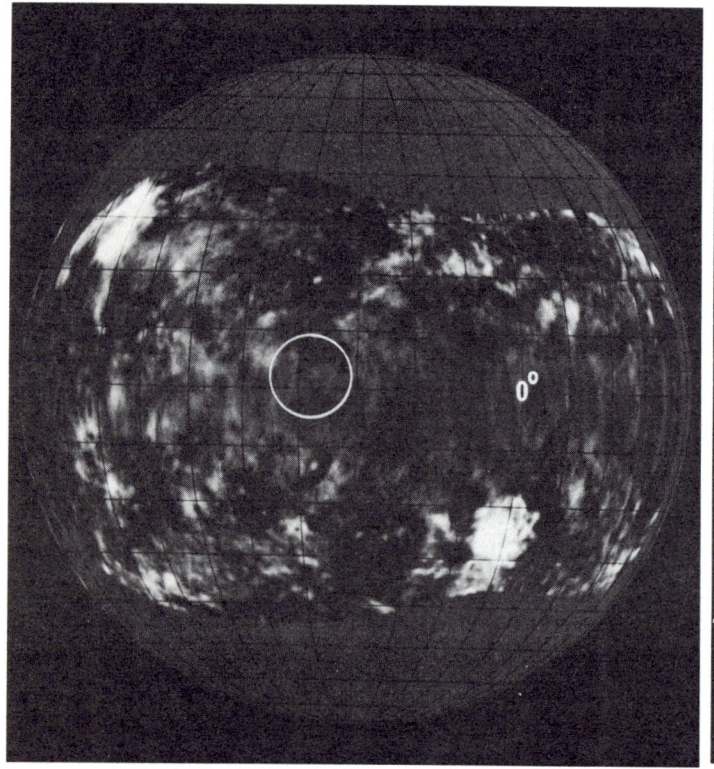

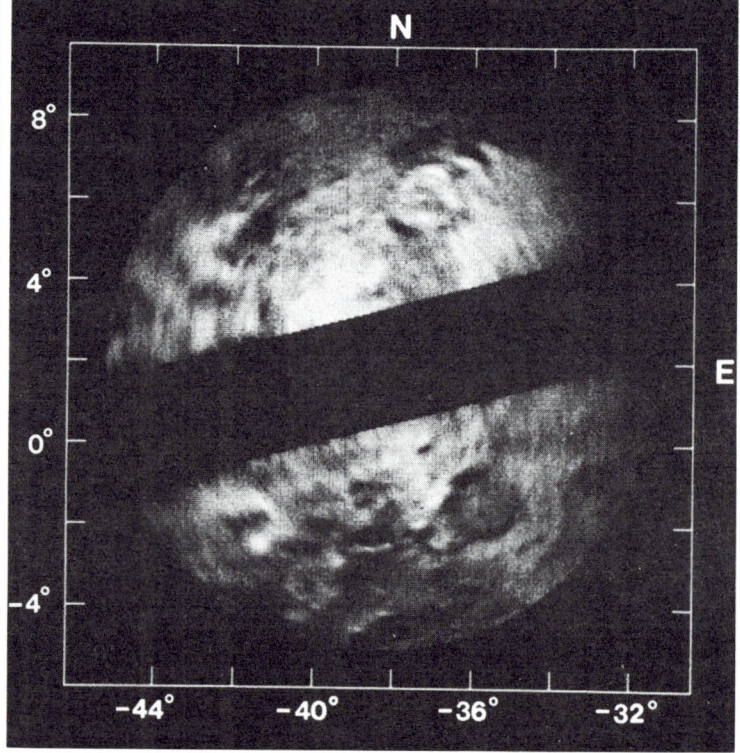

96

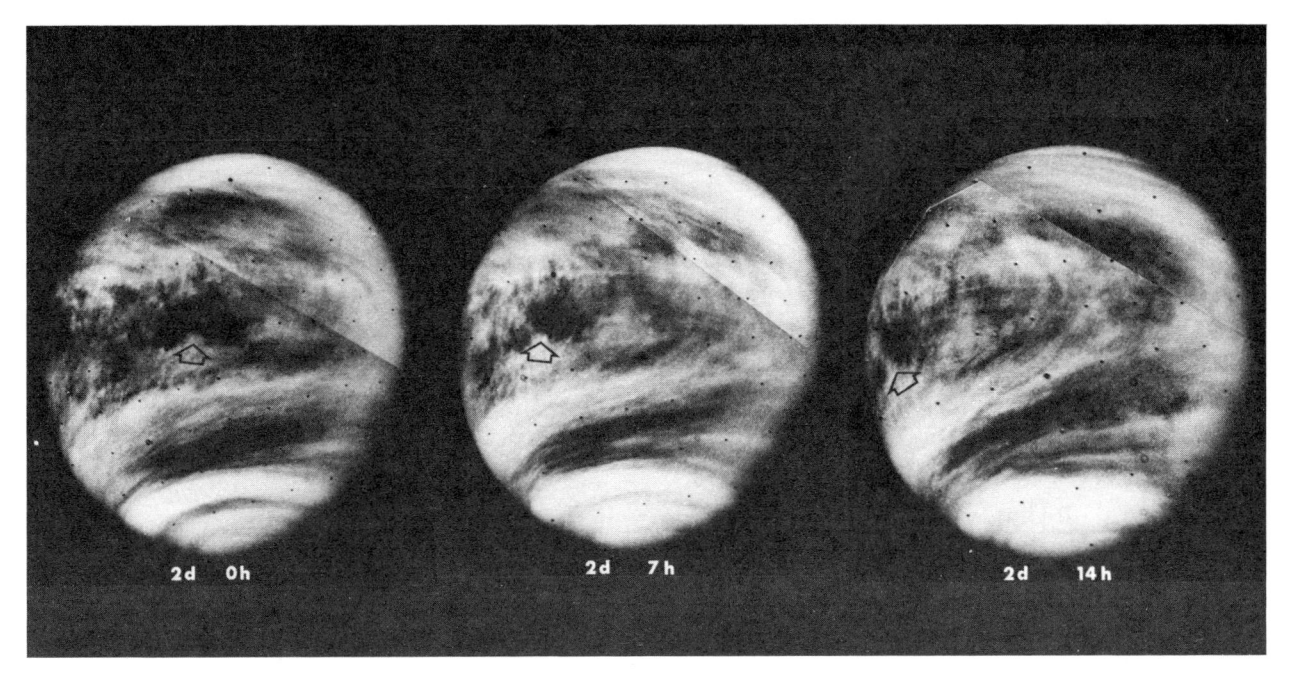

This series of photographs was taken at seven-hour intervals two days after Mariner 10's *closest approach to Venus. Although the planet rotates very slowly, the movement of the cloud masses shows that Venus' atmosphere is in rapid motion. The dark cloud mass indicated by the arrows is approximately 1000 kilometers wide.*

winds, remains a mystery. Perhaps they are not craters formed by impact, but some breed of volcanic crater unknown on Earth.

When *Mariner 10* was designed to fly by Venus on its way to photograph Mercury in 1973 the emphasis was clearly on Mercury. "If *Mariner 10* hadn't carried imaging capability to look at Mercury, we never would've seen the turbulent upper layer of Venus," said Edward Danielson of the Jet Propulsion Laboratory. "We went to NASA and pointed out that ultraviolet light won't go through ordinary glass optics. Still, it was clear that if there was anything to see in the cloud layers of Venus, it would have to be in the ultraviolet. So we made a quartz system, which let the ultraviolet through. That's how we got to see the structure we found."

The structure found was certainly impressive: the top of Venus' atmosphere swirls around and around the planet, repeating a basic pattern every four days. Apparently the driving force in this atmospheric circulation is the sub-solar region—the point at the top of the atmosphere closest to the sun. One would expect this area to be heated the most. At the sub-solar point, which is above Venus' equator, the clouds seem mottled and turbulent. Around this point a curious symmetrical flow pattern has developed. White jet-like streams spiral around the planet and seem to move toward the poles. They are probably wave-like disturbances, and the motion bears a resemblance to the movement of a spiral on a barber's pole when the pole is rotated. Precisely how the heated subsolar point drives atmospheric circulation is not clear. Winds at the equator seem to move about 300 kilometers per hour high above the cloud layer. This may be a sign that slower currents beneath the cloud deck are churning the lower atmosphere. Such currents probably imply that substantial mixing occurs between the cooler outer cloud deck and the warmer interior.

We need to understand the motion of the atmosphere far better than we do, but the outstanding question about Venus is *why* it is so hot. Despite the fact that Venus receives twice as much sunlight as Earth, it reflects the sunlight so well that, without any other factors, it should be colder than the Earth. In recent years the so-called greenhouse model, championed by NASA's S. I. Rasool and C. deBergh, seems to have answered this question at least in principle.

The greenhouse effect helps trap the sun's radiation inside a planetary atmosphere. To begin with, visible light passes through the atmosphere of a planet and warms the surface. This light has relatively short wavelengths. Once heated, the planetary surface reradiates long-wavelength infrared radiation which the atmosphere and clouds will not allow to pass easily. They absorb the infrared, storing the energy as heat. If all the infrared waves were trapped, the planet would continue to heat up. A balance is struck in which some of the infrared waves do escape, stabilizing the temperature. On Earth the surface temperature is maintained about 28°C higher than the Earth would be without the greenhouse effect.

Actually, the greenhouse effect is misnamed. Although the glass walls of a greenhouse do stop infrared radiation, the important heating effect that makes plants grow comes from the fact that the air inside, once warmed, cannot escape. The window panes help by stopping infrared, but that point is not crucial. A greenhouse made of rock-salt panes—that do let infrared radiation escape—would work just as well. In fact, such a greenhouse was built to test this point and it worked perfectly well.

Venus is a logical candidate for the greenhouse effect, because carbon dioxide is quite transparent to visible light but quite opaque at many infrared wavelengths. Carbon

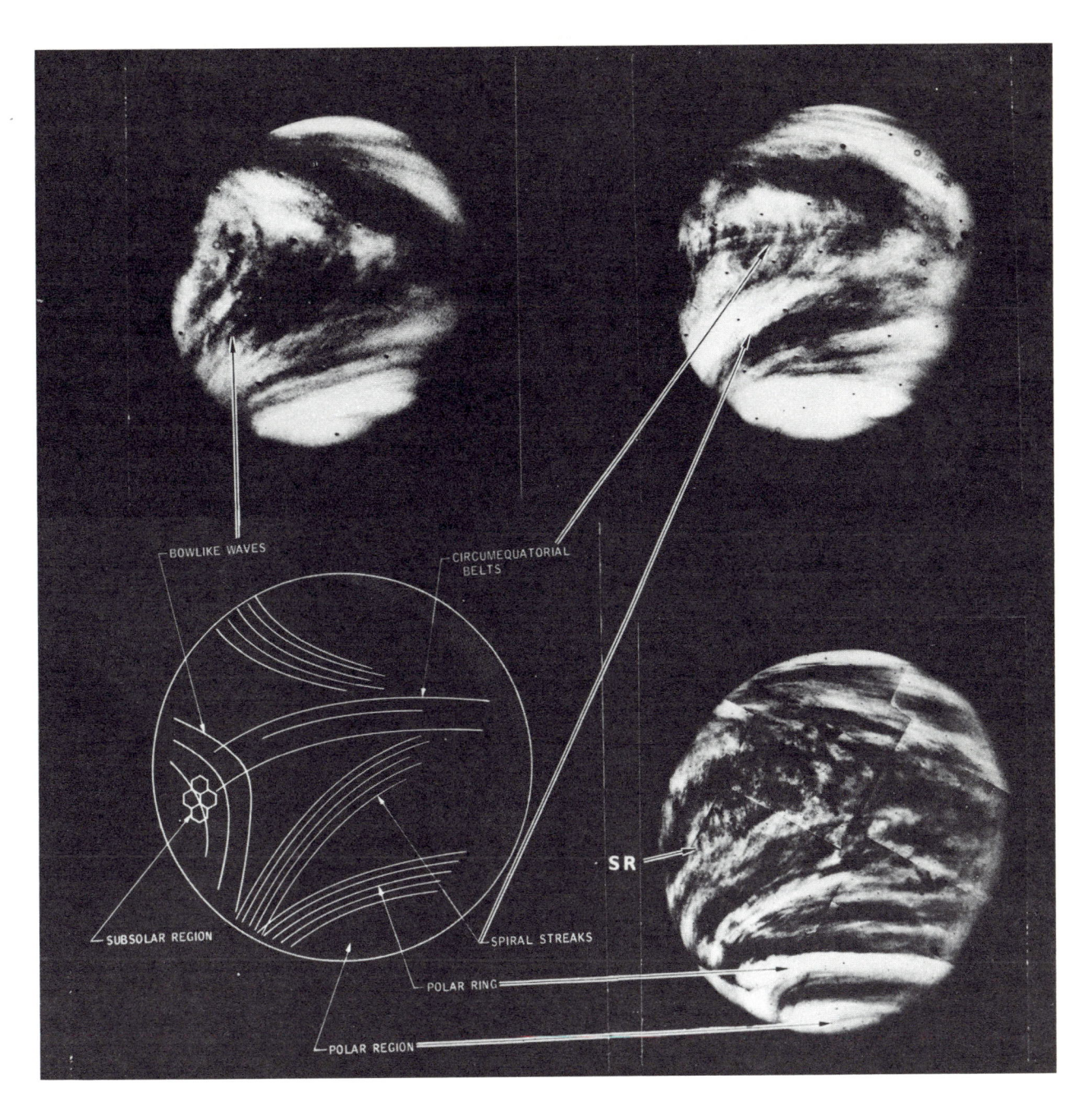

The main atmospheric features of Venus, as seen by Mariner 10's ultraviolet-sensitive cameras. "Subsolar region" is the area directly beneath the sun: high noon on the planet's surface. Apparently heat energy from the sun is the driving force that moves Venus' atmosphere.

dioxide alone, however, could not trap enough incoming sunlight to explain Venus' temperatures. The addition of a small amount of water vapor, quite close to the abundance of percentage reported by the *Venera* spacecraft, would do the job.

That explains why Venus can stay as hot as it is, but how did it get that way in the first place? Rasool and deBergh have made extensive calculations, trying to follow the evolution of Venus' atmosphere. Current thought about the terrestrial-type planets—Mercury, Venus, Earth and Mars—is that atmospheres formed around these bodies when fluids and gases percolating up from the interior emptied onto the surface. Volcanic action supplies liquids and gases such as carbon dioxide, water and other materials at a steady rate. The gravitational fields of the planets held the gases, so that atmospheric pressure increased. What happened next, according to Rasool and deBergh, depended crucially on the ground temperature. Liquid water can only exist in certain ranges of temperature and pressure.

We do not know precisely what the early temperature of Venus was, because we do not know how fast it rotated, that is, how long its day was. We do know that any reasonable duration for the ancient Venus day, given

Carbon dioxide pressure. (Earth atmospheric pressure = 1 on this scale.)

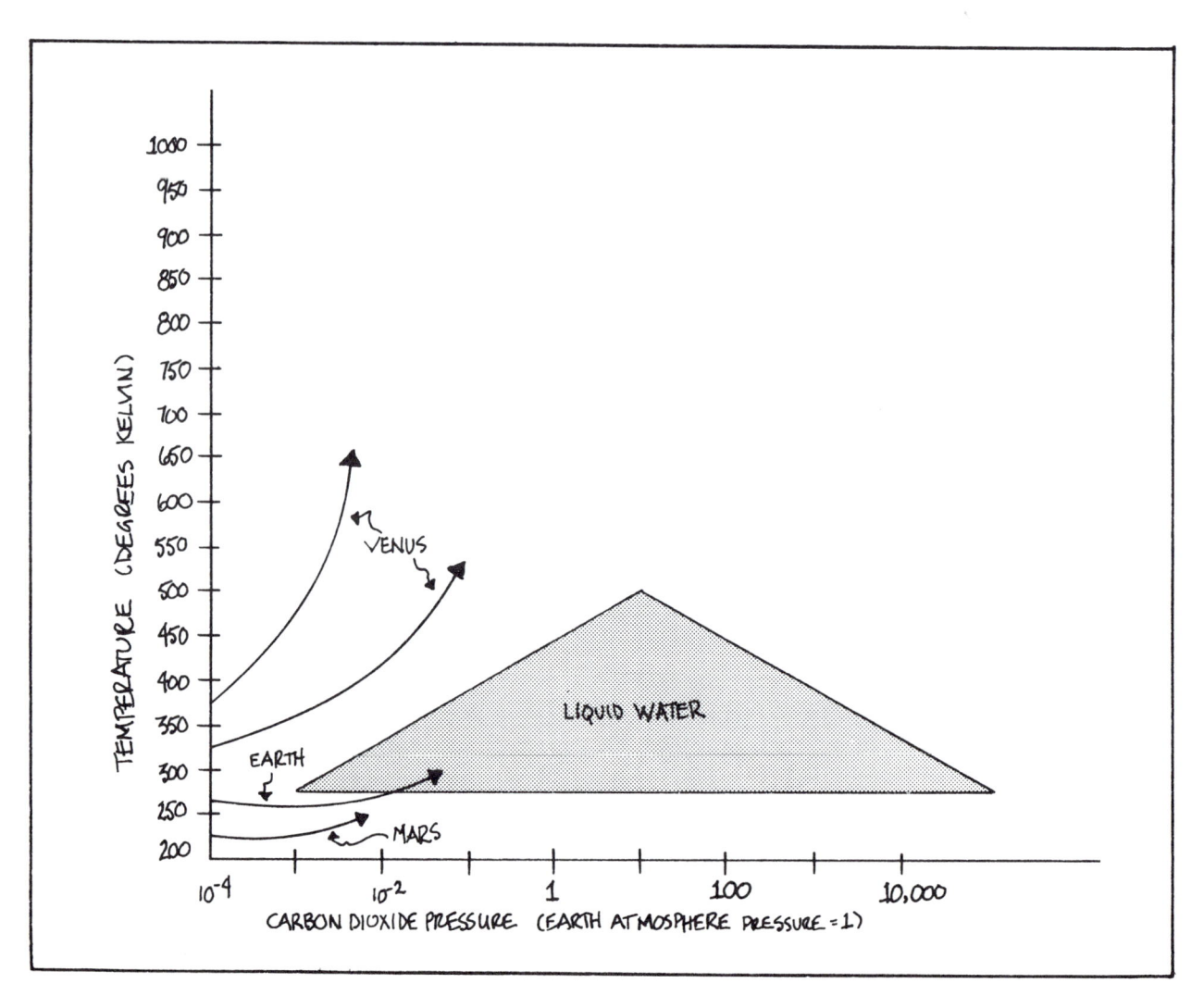

the distance of Venus from the Sun, would have led to a surface temperature above the boiling point of water at the pressures involved. This means that any oceans on Venus would become unstable. As quickly as volcanoes could empty liquid water onto the surface, sunlight would turn it to steam. Water vapor is quite opaque to infrared radiation. A steam atmosphere above Venus would vastly increase the greenhouse effect, heating the oceans further, and thus creating more steam. This is a spiraling effect, a runaway greenhouse.

On Earth liquid water was able to form and remain stable. In the presence of water the carbon dioxide in the air converted into carbonate rocks such as limestone. On Venus there was no liquid water to speed that process, so the carbon dioxide was never trapped into rock. It is still there, in the atmosphere, acting as an insulating blanket sealing in the planet's heat. On Earth carbonate rocks contain enough carbon dioxide to give perhaps 30 to 100 atmospheres of pressure, if the carbon dioxide were released. Thus the total concentration of carbon dioxide in both the atmosphere and the rocks is the same for both the Earth and for Venus. In this sense the planets are quite alike. Venus simply retains most or all of its carbon dioxide in its atmosphere.

Where Earth and Venus differ significantly is in the amount of water they retain. There is at least a thousand times more water on the Earth than on the surface and in the atmosphere of Venus. Perhaps water on the surface of the early Venus was broken up by sunlight into hydrogen and oxygen. Hydrogen, being lighter, escaped. Oxygen perhaps combined rapidly with the planetary surface and was trapped into rocks.

Earth, beginning with a considerably lower starting temperature, produced liquid water and trapped the carbon dioxide in its crust. Actually it seems that the Earth only narrowly avoided the runaway greenhouse effect; if it were 7 percent closer to the sun (10,000,000 kilometers), it would have evolved into an inferno like Venus.

On Mars, the initial temperature was probably lower than the freezing point of water, so the volcanic steam froze on the surface. Only carbon dioxide accumulated in the atmosphere. Mars is small and has relatively few volcanoes. Probably these volcanoes have not given off enough gases to build a significant atmosphere so far. (Or if Mars is currently going through a cold period, a great deal of the atmosphere may be temporarily frozen out at the poles.) In the distant future volcanoes may pump enough carbon dioxide and water vapor onto the surface to begin the greenhouse effect in earnest. Mars could then follow an evolutionary path resembling that of the Earth. Perhaps in that far off time liquid water pools will form, carbon dioxide will be trapped into sediments and an atmosphere of mostly nitrogen will come about, and man (if he is still around) might find Mars a comfortable home.

On Venus, then, there are billowy clouds of ice crystals above, a baked barren landscape below—but what lies in between? Could there be water clouds aloft among temperatures close to that on Earth's surface? Nothing could be said until the early 1970's, when detailed study of the clouds of Venus became possible. The American astronomer Andrew T. Young noted that the clouds absorbed one particular wavelength very strongly. He searched for a chemical compound that could explain the absorption feature. Others had tried various candidates to explain the feature—hydrocarbons, dust and oxides, ammonium cloride—but with only partial success. Young discovered that a 75 percent solution of sulphuric acid fit the data exactly. That corrosive mixture of three parts acid to one part water could exist as a mist at the temperatures and pressures of the Venus clouds. Sulphuric acid absorbs water easily, which may explain why ground-based observers have found only traces of water vapor above the clouds. The roiling, churning clouds have sulphuric acid droplets floating among them. There must be gaseous sulphuric acid below. There may even be a high concentration near the surface; the Russian *Venera* probes could have found such acids, but they were not designed to do so.

If there is a zone of moderate temperatures and water vapor between the clouds and the searing surface, rains of sulphuric acid would make it a seemingly difficult environment for life to begin. Whether biological processes could get under way amid these acid-choked clouds is still an unexplored question.

We could imagine some kind of life floating in this thick atmosphere; after all, some bacterial life floats freely in the clouds of Earth. Any Venus cloud creatures would have to be small, light and able to survive in an acid environment. Recently there has been clear evidence for hydrochloric acid and hydrofluoric acid in thin gases among the upper atmosphere. Life would have to deal with all these substances, and probably more. If it could, even then it would have to survive in a relatively narrow layer. Although the atmosphere is thick and the clouds we see are some 60 kilometers above the surface, the ultraviolet photographs of *Mariner 10* show a slow steady churning in this vast blanket of carbon dioxide. That means currents constantly carry material from one level of the atmosphere to another. Only in some regions would the temperature be moderate and pressure high enough for small life spores to float on the updrafts. Something like a bladder fish might live there, negotiating the winds. Or perhaps bacterial life might multiply there so quickly that it could survive its losses. There would always be winds to sweep life down into the scalding depths, or cast it upward into the freezing clouds. It would be hard to imagine a more demanding place for life to arise.

But we know very little about the precise conditions necessary for life, so any conclusions about Venus are premature. To paraphrase D. Deirmendjian of the Rand Corp., commenting on the lower atmosphere of Venus, "We know a little more than we did, much less than we should, and understand much, much less than we know."

The Venus Pioneer Program scheduled for 1977 and 1978 may correct this. *Pioneer* will orbit the shrouded planet, mapping the clouds in visible and ultraviolet light, and searching for clues to the temperature and water vapor distribution deep below the clouds. The question of possible life can only be solved by a direct descent through the enigmatic clouds and into the warmer layers below. *Pioneer* is not designed to enter the atmosphere, but later probes undoubtedly will be. What they find there will influence our ideas of planetary evolution, and may yield even more surprises in the mystery of Venus.

What then? Man is a meddler. Not only does he not leave well enough alone, he constantly tries to improve even the hostile parts of his environment on the Earth. It is natural to think that in the far future, with Earth crowded and overburdened, man might consider other places to live in our Solar System. Mars is an obvious candidate. But surprisingly, there is even a chance that Venus might some day be gently altered until it resembles something like the early Earth.

Terraforming an entire planet is as prodigious a task as we can envision, but it is not absurd. The work of Rasool and deBergh implies that Venus took a wrong turning early in its evolution, following a track our own Earth might have chosen if it had been somewhat closer to the Sun. This is a clue that Venus might even now be diverted back toward more Earth-like conditions.

The crucial ingredient is life. The early atmosphere of Earth was a noxious mixture of methane, ammonia, carbon dioxide and other gases. We owe our present oxygen-rich air to the presence of life. It is possible that suitably engineered simple life forms could perform a similar task for Venus.

Suppose we began by fertilizing the upper atmosphere of Venus with small, buoyant photosynthetic algae. They would have to ride the turbulent currents there, feeding off carbon dioxide and any water present and surviving the choking mists of acid. Their life processes would be driven by photosynthesis, just as on the Earth. These hearty algae could, if designed precisely, convert the surrounding gases into organic substances. They would leave behind a residue of oxygen.

Of course, in the turbulent winds of Venus the algae could not survive forever. We would have to inject them at a steady rate from above. They would inevitably be caught in downdrafts and fried in the lower levels. Heating algae releases carbon, simple carbon compounds and water. Thus the water consumed at the first stage of an algae's life would be returned to the air when it was cooked below. The only net change due to the algae's existence would be a slight increase in the oxygen content.

The oxygen could not survive long. Inevitably it would be chemically combined with the desiccated, baked surface. Gradually matter would thus be extracted from the atmosphere and stored in the crust of Venus. The crushing pressure of the atmosphere would slowly decline, and with it the atmospheric greenhouse effect would weaken. That would lower the temperature of Venus, probably expanding the zone of the atmosphere in which algae could survive. Thus these small creatures could expand their own life site in the clouds, fueling a spiraling process that would convert the atmosphere.

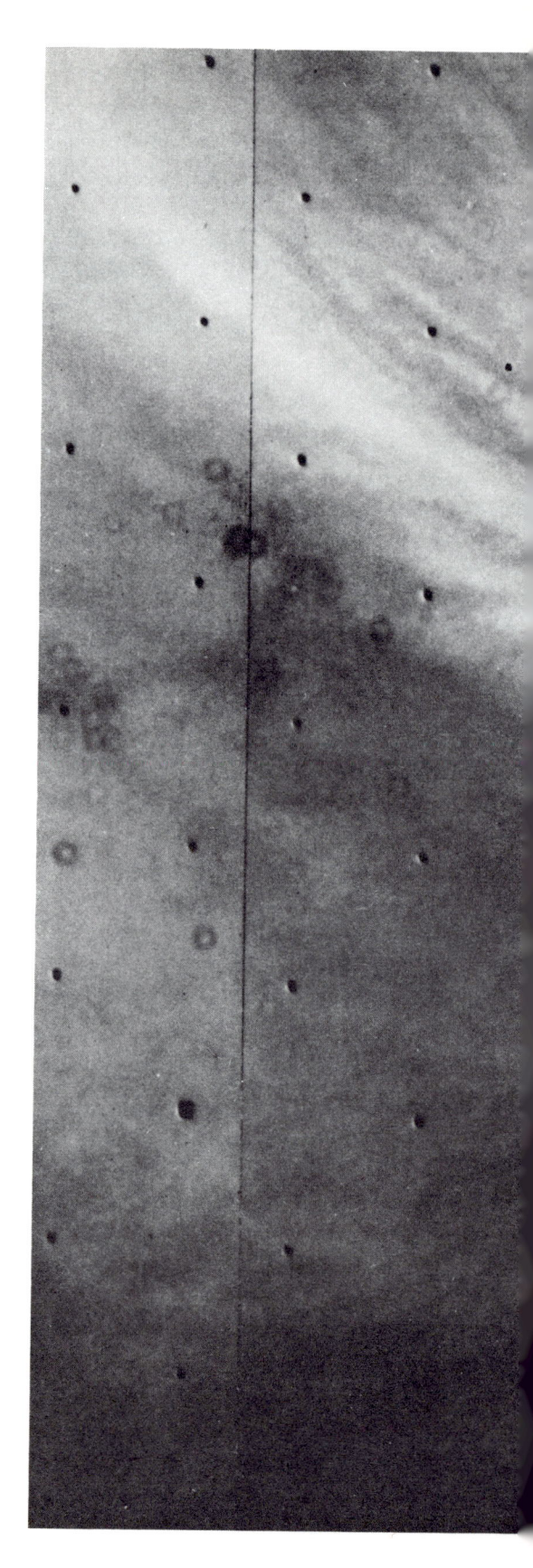

To keep the algae going, we either must inject them rapidly into the upper atmosphere or make them particularly good at reproducing themselves. Either course would give a runaway process that, in time, might convert the incredibly hostile environment into something resembling the early Earth. The water content of the present atmosphere of Venus, if condensed out onto the surface with no losses, would give a shallow pond one foot deep. Not very much, by terrestrial standards, but a beginning. There might be more water locked up in the rocks.

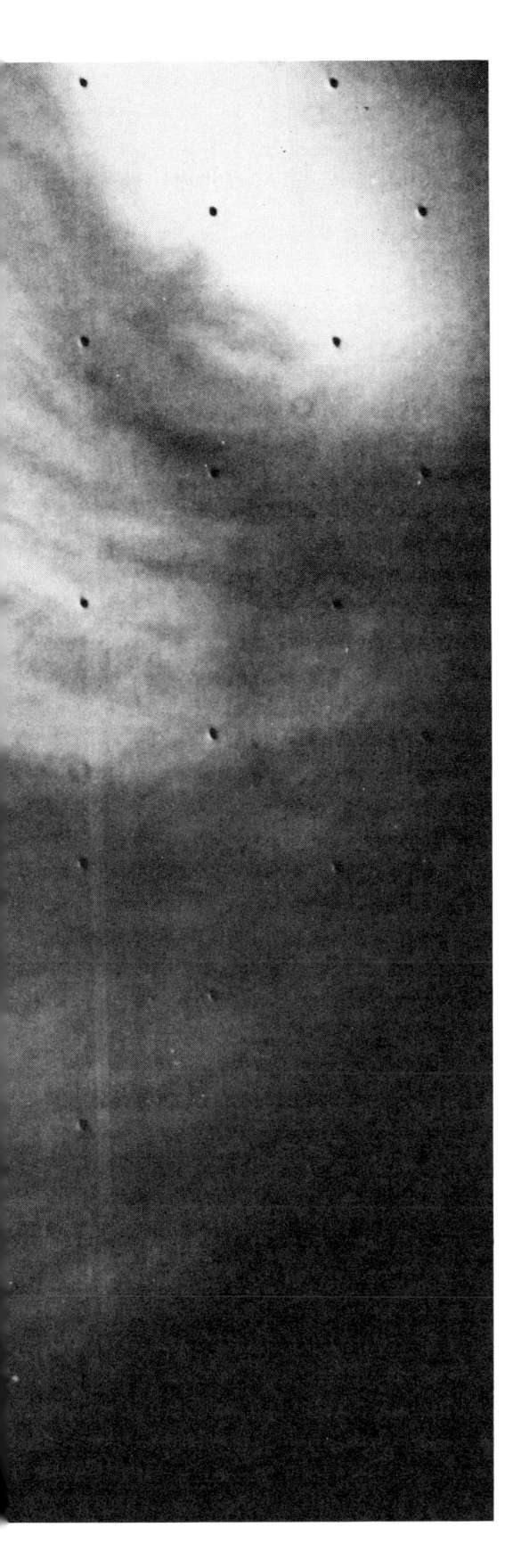

With a subtle, delicate control of the process we might even be able to help create an ozone layer like Earth's above Venus' atmosphere. It would shield out ultraviolet light and allow single-celled green algae to thrive throughout the cooling atmosphere, and perhaps eventually let them colonize the surface itself. At that point wholesale redesigning of the Venus surface could follow. New life forms could be planted and mankind could begin deliberately a process that nature takes literally billions of years to perform.

Of course, this is a grandiose and literally astronomical project that would take centuries to accomplish. But man has come so far in such a short time, and his knowledge is expanding so rapidly that even such improbable feats may not lie beyond his capabilities. In less than one hundred years our knowledge of Venus has gone from a few comfortable illusions to a welter of hard facts with intriguing possibilities. We are just beginning to understand our neighbor, and by no means are all the returns in. We need to understand Venus because it represents a completely different course in planetary evolution. It is so similar to Earth in many ways, yet it is almost improbably hostile.

If we require an example of how two planetary evolutions can begin with nearly the same conditions and end with completely different results, Venus and Earth provide it. Since we are now tinkering with our own atmosphere by pouring smoke, fumes and smog into it, Venus may be a useful object lesson of where we could go wrong.

The southern hemisphere of Venus, from a distance of 720,000 kilometers. The spiral bands of clouds indicate cloud flows that originate in the equatorial regions and move toward the pole.

JOE HALDEMAN was born in Oklahoma in 1943. As a teenaged amateur astronomer in Maryland, he coordinated a worldwide network of other young astronomers during the summer of 1960 who kept a twenty-four hour watch on all the visible planets. Data from his project earned several of the participants Westinghouse scholarships.

After earning a degree in astronomy in 1967, Haldeman was drafted and fought in Vietnam. This experience provided the background for his first novel, *War Year* (Holt, Rinehart and Winston; 1972). He uses his background in physics and astronomy in his short stories, which appear regularly in science-fiction magazines, and he covered the *Apollo* and *Skylab* launches for a Florida newspaper. Of his five other novels, three were adventure potboilers written behind an ironclad pseudonym. The fourth was *The Forever War,* and the fifth *Mindbridge,* science-fiction novels published by St. Martin's Press. He also edited *Cosmic Laughter,* an anthology of humorous science fiction (Holt, Rinehart and Winston; 1973) and has done experimental work in television. He played the town drunk in a grade-B vampire movie.

Currently Mr. Haldeman is teaching English at the University of Iowa, while belatedly working toward a degree in the same subject.

The Surprising World Called Mercury

Joe Haldeman

	MERCURY
mean radius	2446 kilometers
mass†	0.054 kilometers
mean density	5.44 grams/cubic centimeter
surface gravity†	0.37
escape velocity	4.2 kilometers/second
length of day*	58 days 16 hours
length of year*	87.96 days
inclination of orbit to ecliptic	7°.0
inclination of equator to orbit	0°.0
mean distance from Sun	0.387 AU = 5.79 x 10^7 kilometers
eccentricity of orbit	0.206

satellites	diameter (km)	distance from planet (millions of kilometers)	orbital period (days)*	date of discovery
None	NA	NA	NA	NA

† Earth = 1
* Earth time

Over 2500 years ago, the blind poet Homer could enchant his Greek audience with tales of fabulous lands, supposedly only a few hundred miles from where they sat listening. True, no map could tell you where the Land of the Lotus-Eaters was. But there were many uncharted islands. And many ships that never came back.

Homer's Mediterranean was eventually charted too well, and storytellers had to look elsewhere for their fantastic places and peoples. It was no problem: Africa, far Cathay, the mysterious Americas. It was a big world.

But the world kept getting smaller. Your neighbor went to Africa and came back with snapshots. A guy from far Cathay ran the laundry on the corner.

The rugged face of Mercury, photographed by Mariner 10 from a distance of 77,800 kilometers (49,000 miles). Although the surface looks much like the Moon, Mercury shows a greater variety of surface features. In the top center of the photograph, a strong eastward-facing scarp extends from the limb (edge) of the planet southward for hundreds of kilometers.

Well, there was still the rest of the Solar System. You could tell stories about all manner of strange creatures cavorting around on Mars or Saturn, and your audience would go along with it. Nobody really knew what it was like out there.

Slowly but surely, the astronomers closed in on the storytellers. There aren't any butterfly men on Saturn: it's too cold, there's too much gravity, the atmosphere is full of stinking poisons.

Some good storytellers, though, simply accepted what the scientists said, and wrote their stories around the facts. If Mars is cold and its air is very thin, then Martians are barrel-chested, furry creatures. The beasties who live on Jupiter have to contend with crushing pressures and hurricanes of corrosive gas? Then they look like huge armor-plated caterpillars who take baths in liquid methane and complain if the weather isn't bad.

The picture astronomers gave of the planet Mercury was especially handy for storytellers. One scorched side of the planet always faced the Sun; the other side was never warmed by it, and was colder even than far-off Pluto. There was a thin "twilight zone" around the circumference of the planet, where the Sun appeared to slide slowly above and below the horizon. Humans could live in this narrow strip, if they brought their own air and water.

In less than a day's travel from your twilit base, you could venture Sunside to the bleak and hostile desert where cataracts of molten lead splashed down the sides of bleached granite cliffs. Or you could go Darkside, and in the cold light of unblinking stars search the frozen ground for pockets of oxygen, to shovel up like snow and carry back to your precarious haven.

It looked as if the scientists had cooperated with the storytellers, for once. It was a powerful metaphor, men trapped between fire and ice. Unfortunately, the astronomers were wrong. And they stayed wrong, right up to the dawn of the space age.

Mercury is, of course, the closest planet to the Sun. Think for a moment of what this means to an astronomer who is trying to get a good look at the planet.

The greatest possible separation between Mercury and the Sun is less than 28 degrees*. This means that the only times you can see Mercury are sundown and sunup, and it is never farther from the horizon than the distance you can cover with your palm, at arm's length. Even at best, then, it is dimmed by the background twilight, and its light has to travel through the haze and fog near the horizon.

It's probably a safe bet that not one person in a hundred has ever noticed Mercury, lurking so near the Sun at dusk or dawn. Many people who make a concerted effort to find the planet are never able to see it. Copernicus, the 16th-century astronomer who dared to say that Earth was just another planet going around the Sun, supposedly complained on his death-bed that he had never managed to sight the elusive planet.

Even if the astronomer manages to find Mercury and train his telescope on it, he's not likely to see very much. The telescope doesn't just magnify the planet's tiny disk; it also magnifies the Earth's shimmering air, which is always at its worst close to the horizon. And the haze and background light tend to wash out the details of the planet's surface. So at best he can make out a few amorphous smudges.

There are other problems, too. Because it swings in such a tight orbit around the Sun, Mercury just isn't visible most of the time. Under ideal conditions (which would be an observatory on a mountain, near the equator, with 365 cloudless days per year), an astronomer might sight Mercury 70 times per year. He usually would be able to observe it for only a few minutes, and never longer than an hour**.

Astronomers got the fire-and-ice model of Mercury by deducing from observations that the planet rotated on its axis once every 88 days. Since Mercury also takes 88 days to go around the Sun, it follows that half the planet is always facing the Sun, and the other half is constantly in shadow. This equality of revolution and rotation periods

---

*Astronomers measure the apparent distance between heavenly bodies in degrees, minutes, and seconds of arc (there being 60 minutes in a degree and, logically enough, 60 seconds in a minute). The distance from the horizon to the zenith, directly overhead, is 90 degrees. The full Moon is about half a degree wide.

** A good description of the process can be found in *Earth, Moon, and Planets,* by Fred Whipple. Harvard University Press, Cambridge, 1968.

(top) *The face of Mars, as seen by the* Viking 1 *spacecraft a month before its lander touched down on the Martian surface. The mammoth crater near the day-night terminator is Argyre. Part of the "Grand Canyon of Mars," Vallis Marineris, can be seen on the left. Along the right limb (edge) of the planet is the Tharsis region bright with cloud activity.* (right) *The first color photograph from the surface of Mars, taken 21 July 1976, shows the rock-strewn rust-hued soil and dust-laden pink sky of the Red Planet.*

(top) *A working day on Mars. The Viking 1 lander scoops soil samples for the miniaturized biology laboratory aboard the vehicle. The scoop and movable arm are in the upper right of the picture. At lower right corner is the meteorology instrumentation, which made measurements of the Martian weather. The short boom at left contains additional instrumentation.* (far left) *Martian sunrise is captured by camera on Viking 1 lander 26 July 1976, at the beginning of the spacecraft's sixth day on the surface of Mars. The lander craft bears the flag of the United States, the Bicentennial symbol, and the symbol of the Viking program—an ancient Viking longship. To the right is the Reference Test Chart used for correcting the balance of the color pictures.* (center left) *What lies over the horizon? The Martian surface is covered with rocks, some of which appear to be of volcanic origin. The red soil is rich in silicon and iron, and shows signs of both wind and water erosion.*

Jupiter, largest of all the planets, as seen by Pioneer 11 *on 1 December 1974. The spacecraft was 2,100,000 kilometers (1,240,000 miles) from Jupiter. Note the shadow of Io, one of the four large satellites of Jupiter.*

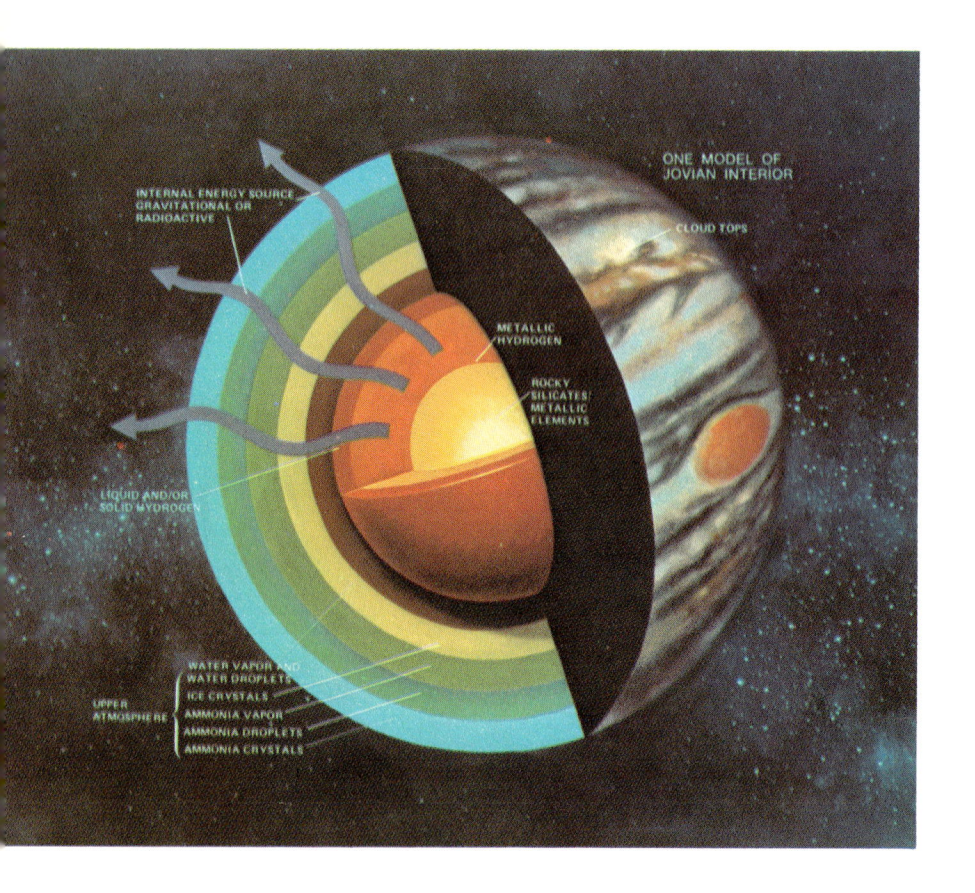

(left) *Beneath Jupiter's clouds is a giant planet that gives off two to three times more energy than it receives from the Sun. Scientists speculate that Jupiter has a core of rock or possibly metallic hydrogen, overlaid by layers of liquids and gases. The source of Jupiter's interior energy production is unknown.* (bottom) *The closest view of Jupiter's Great Red Spot, taken by* Pioneer 11 *when the spacecraft was 545,000 kilometers (338,000 miles) from the Jovian cloud tops. Details of the flow of clouds around the Red Spot, and along the various belts and zones, stand out sharply.*

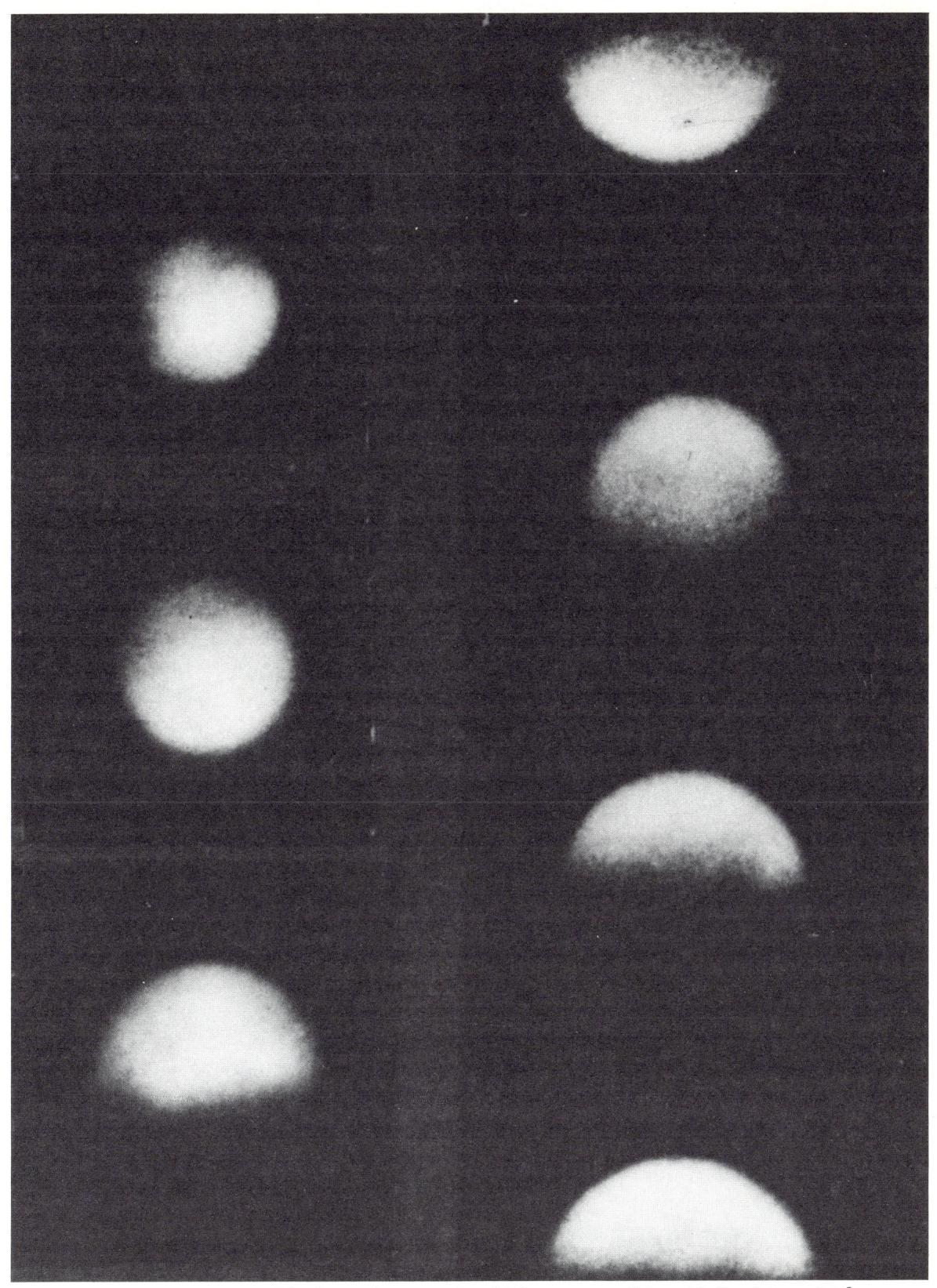

The best photographs of Mercury possible with Earth-based telescopes.

111

is the same phenomenon that keeps the familiar side of the Moon always facing the Earth.

It turns out that the Mercurian "day" is actually 59 Earth-days long. Yet generations of astronomers accepted the 88-day period—one team even claimed to have found the rotation and revolution periods equal to within little more than two hours! How could all of those people have made a consistent 29-day error?

Before we condemn them for having been sloppy observers or conspirators ("Let's give the telescope a rest tonight. Giovanni says it's 88 days and that's good enough for me."), we ought to consider what one has to do in order to determine the length of a planet's day.

Ideally, you look at the planet through the telescope and find some prominent marking. Then you watch the planet rotate, stopwatch in your hand, and time how long it takes for the marking to come around to the same position again. This would work if the planet were cooperative enough to spin around once before it set, but most planets won't do this (it might work with Jupiter, Saturn and Uranus, which can under certain conditions be visible most of the night, and all of which have "days" less than 11 hours long). You usually have to take a long series of timed observations and figure out the period from them.

The Moon rotates once for each revolution, and thus always presents the same face to the Earth. Until recently, astronomers thought that Mercury behaved the same way with respect to the Sun.

A simple demonstration: you look at planet X on Tuesday night and record the fact that a purple dot is right in the middle of it, at midnight. The next two nights are cloudy. Friday night, at midnight, the purple dot is in the same place. Does this mean that the planet rotates once every three days? Of course not. It could have a 24-hour period, and have gone around exactly three times. Or six times with a 12-hour period. After a number of observations, though, the truth will emerge*.

Now imagine trying to make this method work with Mercury, which has only vague markings, and which you can only observe for short periods of time, under difficult conditions. Even though astronomers had been looking at Mercury since the invention of the telescope (around

---

*Actually, it's more complicated than this, as the data are never so neat. It takes a lot of statistical paperwork to arrive at the answer.

1609), nobody recorded having seen any details on the surface until the early nineteenth century. The two German astronomers who first saw these markings (K. L. Harding and J. H. Schröter) noted that they always seemed to be in the same position, day after day. The mathematician Bessel deduced from their drawings that Mercury's rotation period was 24 hours and 35 seconds.

The 88-day story begins in 1882, with the Italian astronomer Giovanni Schiaparelli. Schiaparelli brought an impressive reputation to the Mercury problem: his painstaking, dramatic drawings of the planet Mars were generally thought to be the best ever made. Even the general public was aware of his work.

Schiaparelli observed Mercury for seven years—and not just in brief twilight glimpses, when the planet is visible to the naked eye. His instruments were powerful enough to make observations of Mercury during broad daylight, so he could track a given detail for hours on end.

Thus he saw immediately that the 24-hour period must be wrong. In all of his observations, he saw no evidence of rotation whatsoever. So either the planet did not rotate (which was very unlikely), or it had a very long period.

Astronomers in Schiaparelli's day had computed reliable rotation periods for every planet from Earth to Uranus*, and at least one asteroid as well. None of them had a period much longer than 24 hours—with the single exception of Earth's Moon, lockstepping around the Earth the way we described earlier.

Scientists thought that the Moon had gotten its synchronous period because of tidal friction, a result of its having been close to the Earth while still in a molten state. The pull of the Earth had made it slightly egg-shaped, and it could be shown mathematically that this would cause its period of rotation to slow down until it matched the period of revolution, even if the Moon were out-of-round by as little as a few hundred feet. So Schiaparelli's logic in choosing 88 days rather than, say, 80 or 1,000, seems clear. Mercury was acting with respect to the Sun in exactly the same way as the Moon behaves with respect to the Earth.

There were a few detractors, but for the most part Schiaparelli's observations stood unchallenged for 75 years. As late as 1963, a book about the planet could say "...there can no longer be any doubt that the day on

---

*Neptune had been discovered, but no details could be seen on its small disk.

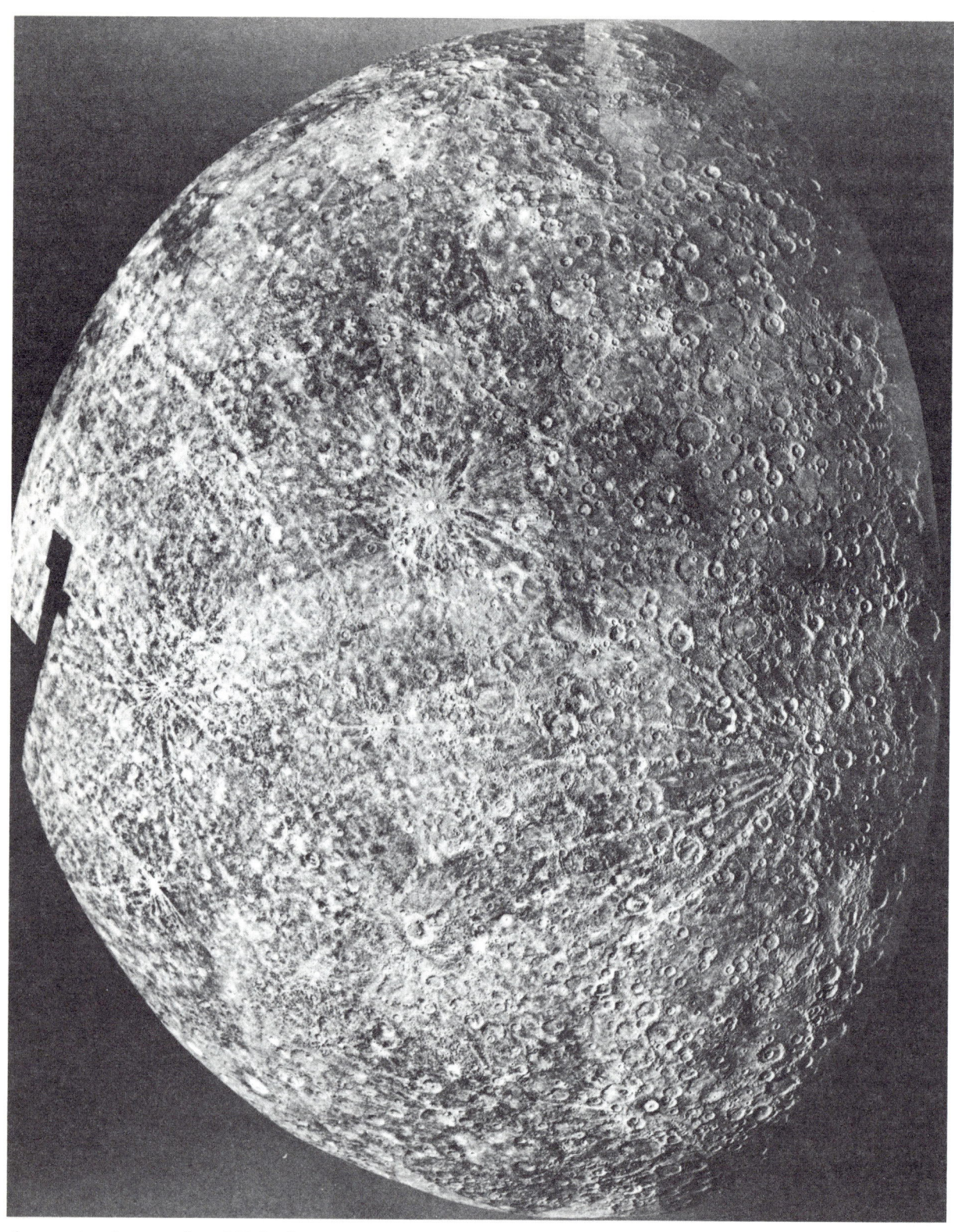

A mosaic of more than 200 high-resolution photos from Mariner 10 shows Mercury's south pole from a distance of about 50,000 kilometers (31,000 miles). The largest craters are more than 150 kilometers across.

Mercury is equal to its year." By 1965, radar observations had proven otherwise.

New telescopic observations confirmed the radar results. Why hadn't somebody done it sooner?

One reason is purely practical. Before space flight, planetary observation had been astronomy's unwanted stepchild for more than thirty years. Large telescopes simply aren't designed to be used on planets: they are essentially huge cameras that can be put to better use in galactic and extragalactic research. Photography is wonderful for stars and nebulae, long exposures revealing things that the human eye could never see, but an Earth-based photograph of a planet is always blurred because Earth's atmosphere is never still for long enough. Visual observation of planets is far superior, but takes too much time, with one patient astronomer peering at a planet for hours, his brain assembling a picture of the planet from what he sees during split seconds of atmospheric clarity. And telescope time is worth too much to be used so inefficiently.

Two other reasons are less important, perhaps, but also less easily defended.

Giovanni Schiaparelli was one of the greatest astronomical observers of his time, when astronomical photography was in its infancy. Visual acuity was still an astronomer's most important tool. So it's understandable in simply human terms, that most astronomers would be reluctant to invest years in verifying Schiaparelli's observations, when they could be working on something new; something entirely their own. And what if your observations contradicted the synchronous-rotation theory? Schiaparelli had already proven with his observations of the Martian *canali* that he could see things other observers missed. Much of what he saw was optical illusion, but that would not be proven for many decades.

The other problem is rather more subtle. A theory existed that explained why the Moon always presented one face to the Earth. But the Moon itself was the only illustration of the theory. When Mercury's rotation also seemed to show the effect of tidal friction, it seemed *right;* it was consistent with what we already knew about the Solar System.

This is not a case of "why rock the boat?"—the best scientists delight in rocking boats—but only that it seemed

115

persuasively logical and elegant. There was no real reason to go gunning for Schiaparelli.

Astronomy used to be a comparatively passive science. A geologist, for instance, can pick up his rock and hit it with a hammer. A zoologist can vary an animal's diet, or see how it reacts to a change of surroundings or company. A chemist can mix his chemicals together according to any whim; he can check their density, solubility, and even taste (if he lacks prudence).

But before the space age, astronomers for the most part had to be content to sit back and simply look at their objects*. All you can do with a star is chart its position, record how bright it is, and analyze its light with a spectroscope. If what you're looking at is a planet, you could also draw a telescopic map. But you couldn't reach out and touch it. Not until radar was invented.

The theory of radar is based on the simple fact that radio waves move at a set speed, about 186,000 miles per

*The ones who studied meteorites were an exception. But they were really just a hybrid race, you might say: half astronomer and half geologist.

The planet Mercury, as photographed by Mariner 10.

second, and will be reflected by any material object. You can tell how far away an object is by aiming a radio pulse at it, and timing how long it takes for the pulse to return. For instance, if you aim it at an airplane and the pulse comes back one-ten-thousandth of a second later, then you know that the radio waves traveled 18.6 miles. So the airplane is 9.3 miles away, since the waves had to get there and back.

Radar was invented in the 1930's, but wasn't immediately applicable to astronomy. Detectors simply weren't sensitive enough to detect an echo from so far away.

The strength of a radar echo follows an "inverse fourth power law." That means that as the distance between a radar transmitter and its target increases, the strength of the returning signal decreases in proportion to the fourth power of the increase in distance. To put it more concretely, the signal you get back from an object twenty miles away is only one-sixteenth as strong as is a signal returned from ten miles away. And a signal from the Moon, the closest astronomical body, would be only .00000000000003 (3×10^{-14}) as strong.

It wasn't until the 1960's that equipment was sensitive enough for radar astronomy. Astronomers turned their huge dish-shaped antennas toward the Moon, the Sun, and several planets, measuring their distances with an accuracy far greater than was ever before possible.

This was an important breakthrough, but if radar could only be used to measure distances, its usefulness would be limited. Fortunately, it can also measure velocities with great precision.

If you stand by a freeway and listen to the cars pass, you can hear what is called the Doppler effect. The noise of the car's engine (or of its tires on the road, if it has a quiet engine) drops abruptly in pitch the instant the car passes in front of you. That is because sound waves move at a fixed speed in air: if the thing that's causing sound waves is moving toward you, the waves have to squeeze up closer together than they would be if it were standing still. The closer together sound waves are, the higher is their frequency, and higher frequency means higher pitch. Moving away from you, the source makes a lower-pitched sound than it would if it were standing still, because the waves are stretched out.

As a matter of fact, you can measure the speed of the car if you had some way of measuring how much the pitch changed as it went by. Police, lurking at the side of a road, measure a car's speed by the Doppler effect, but they don't use the car's sound. Instead, they bounce a radar pulse off it. The change in the return frequency of the pulse enables them to measure the car's speed with accuracy.

You can do the same thing with astronomical objects, but it takes more power and a more sensitive detector. Thus you can measure a planet's orbital speed or, more germane to our interest in Mercury, the rate of its rotation.

This is quite an observational feat, considering how tiny a target Mercury is. The disk it presents to Earth has about 1/40,000th the area of the full Moon.

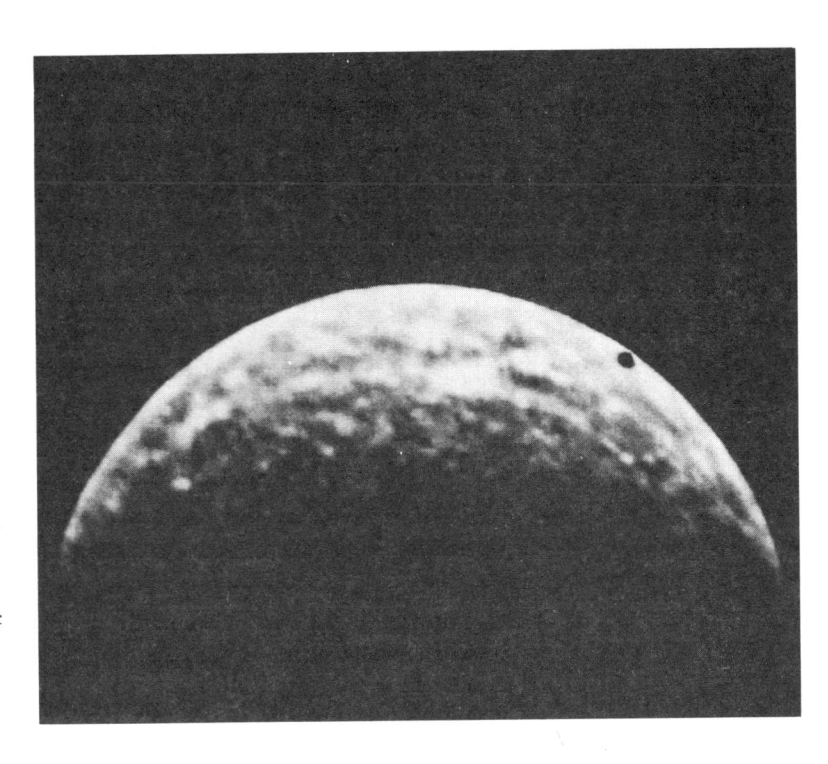

How image-enhancement brings out more information from spacecraft photographs. The smaller picture shows how a typical Mariner 10 *photo was received at the Jet Propulsion Laboratory. After computer-directed image enhancement processing, a wealth of detail of the planet's surface becomes visible. The dark spot is a calibrating mark from the camera.*

119

Radar revealed that Mercury's rotation period was on the order of 60 days, rather than 88. At first this was puzzling, because tidal friction should have slowed the planet down to the 88-day figure long ago. If it were still in the process of slowing down, that meant Mercury simply hadn't been in its present orbit for very long: 400,000 years at the outside.

After making more observations, though, and refining the data, they arrived at the figure of 58.65 days. This is almost exactly 2/3rds the length of Mercury's 87.97-day year. That couldn't be a coincidence. *Something* was happening, but it wasn't as simple as the synchronous lock between the Earth and the Moon.

The explanation for Mercury's odd behavior is in the shape of its orbit. Like all planets, Mercury goes around the Sun in an elliptical path. Most of the other planets, Earth included, have orbits that are very nearly circular. But Mercury's orbit is extremely elongated, squashed.

One of Kepler's laws tell us that a planet's orbital speed increases, the closer the planet gets to the Sun. There is a 24,000,000-kilometer difference between the point in Mercury's orbit that is closest to the Sun (perihelion) and the point that is farthest away (aphelion). Since the average distance is 57,536,000 kilometers, it's easy to see that the Sun has a much more powerful effect on Mercury when the planet is near its perihelion.

This is especially true when it comes to tidal friction. Tidal forces vary as the inverse cube of the distance, so the effect on Mercury is 3½ times greater at perihelion than at aphelion.

The end result is that Mercury does keep one face to the Sun, when it's nearest the Sun. But then as it swings out toward aphelion, the planet continues to rotate at the perihelion-rate, getting more and more out of sync.

This makes for a very odd sunrise-sunset pattern. Suppose you were on Mercury's equator (poor devil!), at a spot where the Sun is just rising as the planet approaches perihelion. The Sun crawls up over the horizon and stays there, motionless, for a day or two. Then it sets. A couple of days later, it rises again, and slowly starts to move up into the sky. As it inches higher, the Sun appears to shrink, and moves a little faster every day, until it gets directly overhead. This is at aphelion, 44 days later. Then the process repeats itself, in reverse order: the Sun, growing, drops to the western horizon, moving slower and slower.

It sets, bobs up again, then sets again. Daytime lasts exactly one Mercurian year; the next sunrise won't be for another year.

The radar observations of Mercury were finished in 1965. At that time, space probes had already visited Venus and Mars, with striking results. It could not be too many years before a probe would go to Mercury, and observe first-hand what radio astronomers had to deduce through painstaking inference from thousands of observations.

The Mercury experiment required the largest dish antenna in the world, the 1,000-foot instrument at Arecibo, Puerto Rico. Why was its precious time taken up with a series of observations that were sure to be confirmed or denied in a few years by a space probe's cameras? Why not simply wait? If the Mercury probe didn't work, or wasn't funded, there would be plenty of time to make the radar observations.

In fact, this illustrates one payoff from space flight that people don't usually consider. The information we get from space probes is invaluable, by itself. But it also provides an objective check of the theories and techniques astronomers used on the planet before the probe furnished close-up data. It is not simply redundant effort, because similar theories and techniques will be used to investigate phenomena that may never be observed close up.

Mercury, in fact, provides an example of this. The very first radar observations of the planet, made in 1963, seemed to confirm Schiaparelli's 88-day period. If the scientists had been content to let these results stand, they would have been in for a real shock eleven years later.

Astronomers had some idea of what Mercury was going to look like long before the *Mariner 10* flyby in March of 1974. It would look like the Moon.

A reasonable guess, you might say, since Mars looked something like the Moon; even Venus, radar told us, has craters. And the chances seemed even better for Mercury: it is a world not much bigger than the Moon in the first place, with no appreciable atmosphere to slow down meteorites or erode away the signs of their impacts.

Further evidence that Mercury was Moon-like was provided by studies of its reflectivity of sunlight. Both bodies were about the same color, and both of them reflected with fair efficiency so long as sunlight struck them head-on, but with poor efficiency if the light struck at an angle. Both bodies have an albedo, or overall reflective efficiency, of 0.07, meaning that only seven percent of the light that falls on their surfaces is reflected. Albedos of the other planets range from 0.15 for Mars to 0.85 for Venus.

This is not to say that nobody expected any surprises, that Mercury would just turn out to be a larger, hotter version of the Moon. As one of the *Mariner 10* scientists pointed out during a press conference before the flyby, the history of space flight shows that surprises are the rule in planetary exploration. Nobody expected Mars to look like the Moon and yet show signs of water erosion; the pressure on the surface of Venus was so unexpectedly high that it crushed the Soviet *Venera 4* probe.

The *Mariner 10* was a cosmic billiard shot that got two planets for the price of one. It was aimed initially at Venus, but in such a way that as it skimmed past that planet

(below) *Layout of the* Mariner 10 *spacecraft, which flew past Venus and Mercury.* (facing page, top) *Mercury as seen from a distance of 1,840,000 kilometers (1,141,000 miles). Craters as small as 160 kilometers across can be seen.* (facing page, bottom) *Mariner 10's double-planet mission flight plan.*

MARINER VENUS-MERCURY

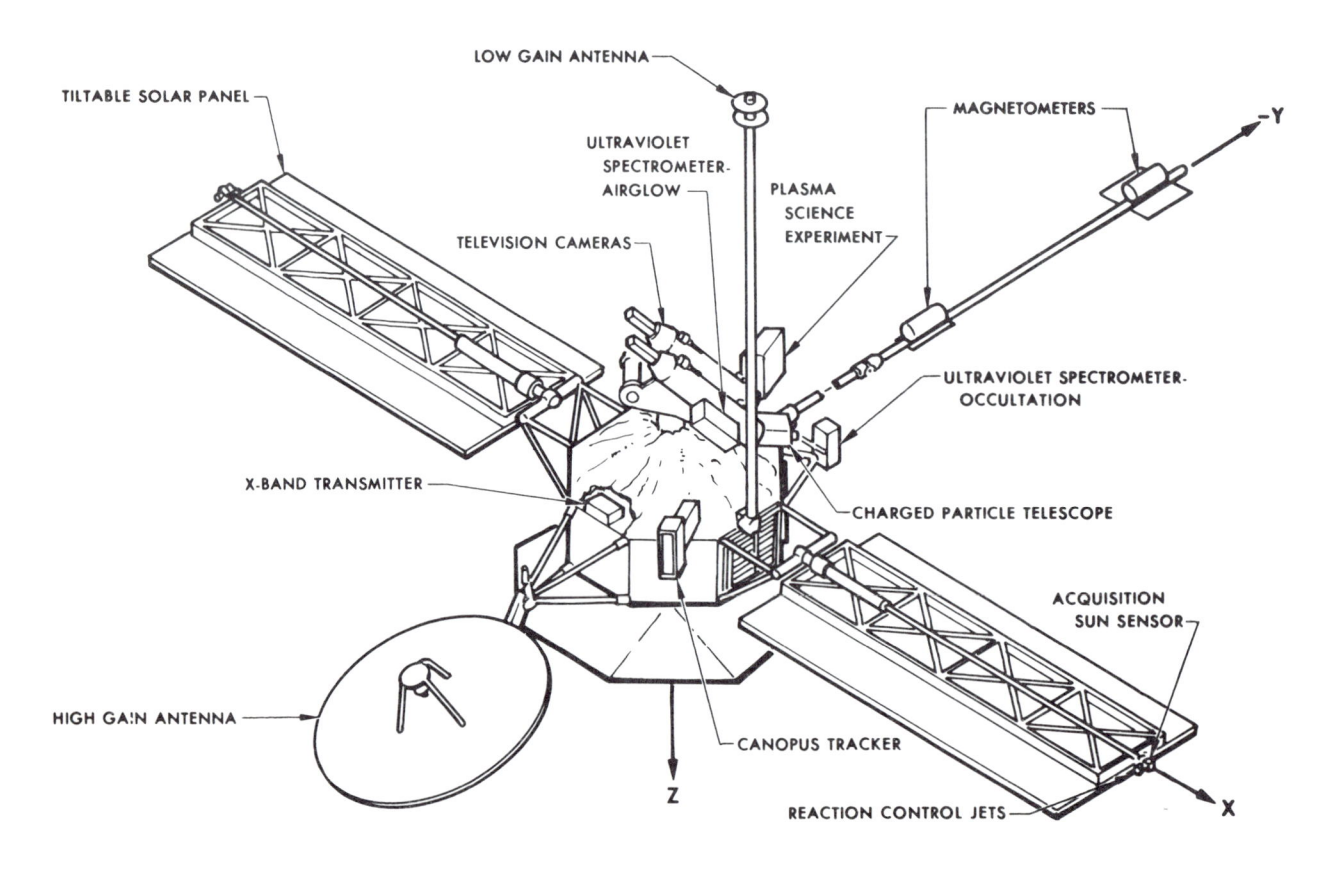

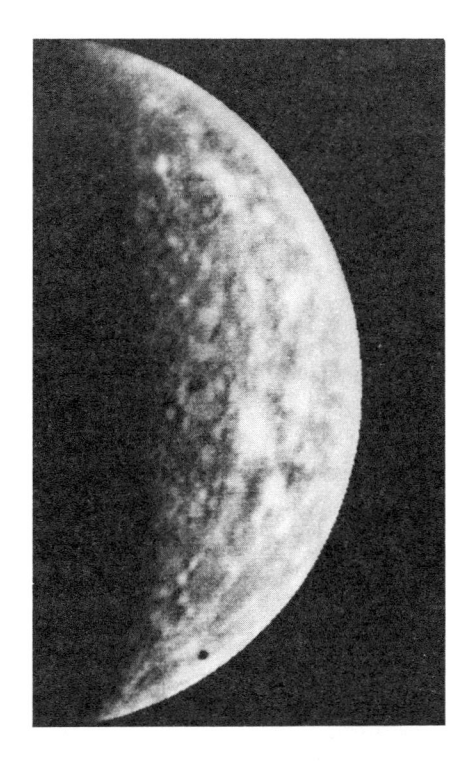

it would be deflected into an orbit that would bring it close to Mercury. It was the first time this technique had been attempted, and it worked perfectly.

The probe came within 696 kilometers of Mercury's surface, taking more than 2,000 pictures. Then it swung around the Sun in a long ellipse and returned to Mercury 176 days later, and collected some 500 more pictures (only coming within 48,000 kilometers the second time). These two passes together provided a view of 37 percent of Mercury's surface. It does strongly resemble the Moon, but it is different in some important respects.

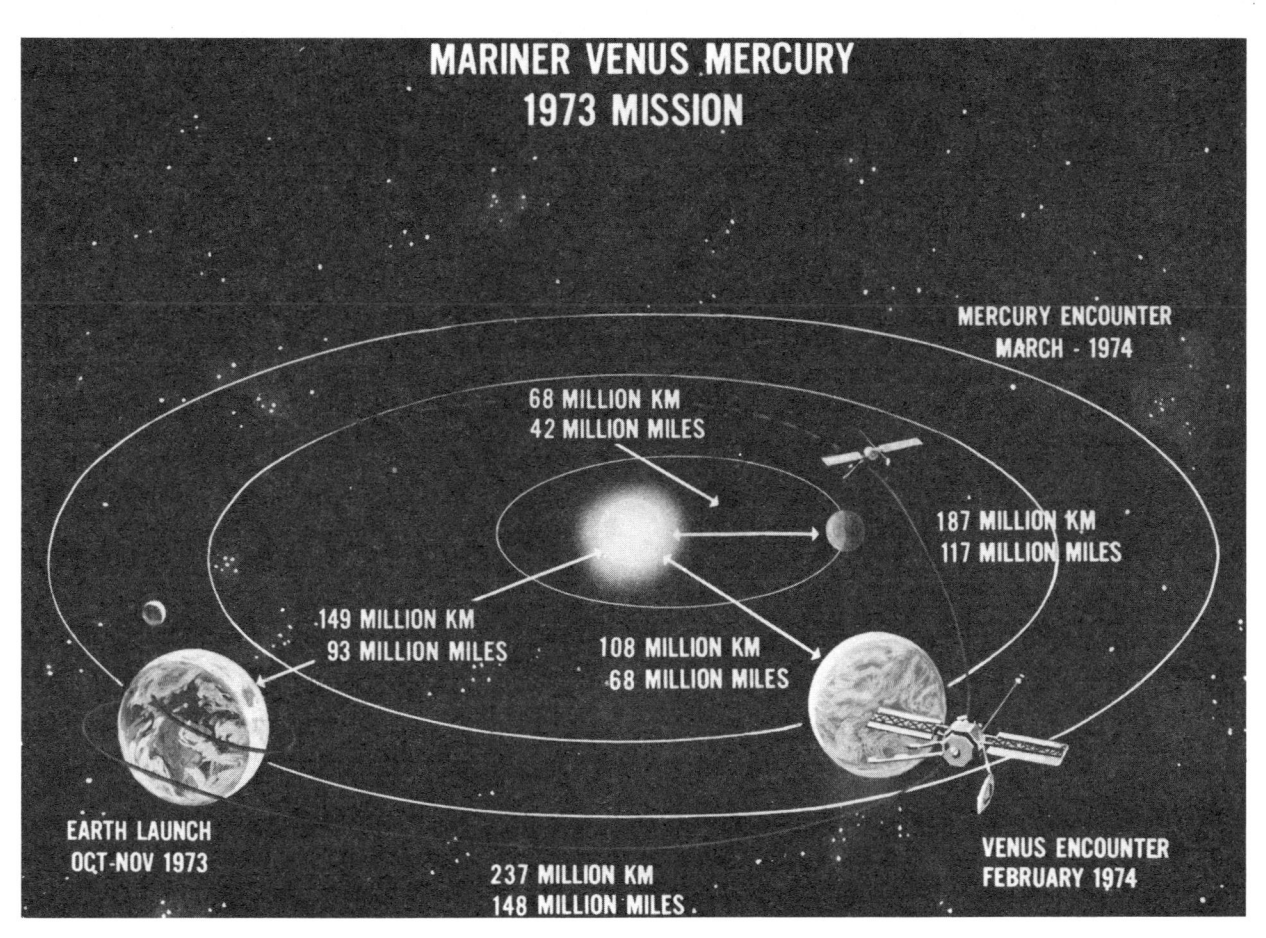

Scientists are being very cautious about drawing conclusions from *Mariner 10's* Mercury data. After all, 99 percent of what we know about the planet was unknown a few months earlier. And most of the data from *Mariner 10* are still "raw data," vast quantities of numbers that have yet to be reduced to meaningful form (each television picture, for instance, was broadcast to Earth as a sequence of 582,400 separate numbers). And almost 2/3rds of the surface has not yet been seen. The lesson of the Moon is a lesson well taken: after centuries of observing the near side—because of libration, more than half of the Moon—the few crude snapshots of the far side that Luna 3 sent back in 1959 completely changed our ideas about lunar topology.

But photography of the surface was by no means the only function of *Mariner 10*. As fascinating as the pictures are, they can't give any information about the temperature of the planet, for instance, or its magnetic field or atmosphere (or lack of either). To measure these and other quantities, the spacecraft was made into a compact, automated laboratory.

The infrared radiometer measured the temperature of various parts of the surface by monitoring the intensity of

A scarp, or cliff, more than 300 kilometers (185 miles) long, photographed by Mariner 10 *while the spacecraft was 64,500 kilometers (40,000 miles) from Mercury's surface.*

A dark, smooth, relatively uncratered area of Mercury, photographed by Mariner 10 *from a distance of 86,800 kilometers (54,000 miles). Much of the area seems to have been filled in with lava flow, which has since solidified, similarly to the* mare *on the Moon.*

infrared radiation from an area "beneath" the spacecraft, as it moved from light side into dark side, and back into the light. It measured an afternoon temperature of 460° Kelvin (368°F.), that fell rapidly to below 150°K. (-189°F.) as soon as darkness fell. As night wore on, the surface temperature fell slowly but steadily to 90°K. (-297°F.), just before dawn, where the surface had endured 88 days of darkness. This is within a fraction of a degree of what scientists call the "oxygen point": at this temperature, if Mercury had an atmosphere, the oxygen could condense out of it and fall like dew.

By studying the way Mercury's surface absorbs heat during the day and loses heat during the night, we will be able to deduce certain things about the composition of the surface.

Three separate experiments searched for a Mercurian atmosphere. The first one was simply to analyze the radio signals from the spacecraft as it disappeared behind the planet's edge, and reappeared on the other side. An atmosphere as tenuous as 1/100,000th that of Earth would have had a measurable effect, but none was present, as had been expected.

The other two experiments were more sensitive, and did show results. One was an airglow instrument, that detected a faint aurora on the planet's night side, indicating the existence of certain noble gases. The other was an ultraviolet spectrometer*, that analyzed the light of stars as they disappeared behind the planet's edge, or reappeared on the other side. The atmosphere these two instruments detected was mostly helium, and amounted to less than one-trillionth the density of Earth's atmosphere.

As tenuous as this atmosphere is, it's far more than Mercury ought to have. Since Mercury has relatively little gravity and is so hot, helium should boil away from its atmosphere in a few hours. Experimenters can explain the presence of helium in two ways. One is that helium is captured from the solar wind, and the other that it is a byproduct of radioactive decay of potassium and thorium in Mercury's crust (this is the main source of helium in Earth's atmosphere). Whether the helium comes from one source or the other, or a mixture of both, they cannot yet say.

Another surprise was the discovery that Mercury has a magnetic field. The Earth's magnetic field is a result of the rapid rotation of its iron core, and although Mercury also has an iron core, the planet doesn't rotate fast enough for this to produce a magnetic field as strong as the one detected (about 100 times smaller than that of Earth). The origin of the field is still a mystery.

Since the radio signals from the other experiments allowed scientists to measure very precisely the position and velocity of the spacecraft, they could apply the well-established laws of satellite motion, and calculate an

---

*At first it looked as if the ultraviolet spectrometer had discovered a small satellite of Mercury, but later observations showed that the object was just another star. It had earlier been theorized that Mercury could not retain a satellite for long; it would soon be pulled away and go into an independent orbit around the Sun.

extremely accurate value for the mass of Mercury. Before *Mariner 10,* Mercury's mass was known to within an accuracy of plus or minus one-half of one percent. After the flyby, the mass of Mercury was known to within better than five one-thousandths of a percent, a hundred-fold improvement.

But it is the pictures of Mercury's surface that most fire the imagination, and will ultimately provide the most important data. What, precisely, do they tell us? So far, scientists can only make cautious generalizations.

That Mercury is very much like the Moon is obvious to any layman. Both have craters and mountains, dark maria and bright ray systems. Mercury presents a less dramatic appearance, as the difference in albedo between the maria and the lighter "highland" areas is not as great as on the Moon. And there are no large **maria,** such as the ones that make up the familiar face of the Man in the Moon.

Also, Mercury seems to have relatively few large circular basins. These are lava-flooded craters caused by the impact of a very large body (tens of kilometers in diameter). They may mostly be on the side we haven't seen, of course.

Mercurian craters are strikingly similar to lunar ones, in size and shape. They seem rather shallower, and the material blasted out of them didn't fly as far as on the Moon, but this is logical, considering that Mercury's gravitational pull is more than twice that of the Moon. From this similarity it is probably safe to assume that the craters on both worlds were formed in the same way, and that the same sort of erosion mechanisms have been at work since they were formed. Lunar exploration revealed that some 90 percent of the craters on the Moon were formed during a period of intense bombardment, more than four billion years ago. A similar period probably accounts for most of the craters on Mercury and Mars.

One type of feature that is rather common on Mercury has no counterpart on the Moon or on Mars. These are the large scarps, which are often more than 500 kilometers long, and can be more than three kilometers high.

Scientists think that the scarps may be "reverse faults," caused by the planet's having shrunk slightly in its infancy. It looks as if the scarps were formed at about the same time as the craters, because sometimes a scarp will run right through a crater, while at other times the crater will interrupt the scarp. If this is truly how, and when, the scarps were formed, then they provide valuable insight into the composition of Mercury's interior.

Scientists long ago were able to calculate Mercury's mass with some accuracy, as we mentioned, by observing its effect on the orbit of Venus. Also knowing the planet's diameter, they could compute the density: 5.5 times the density of water. The Earth's density is 5.52, and the Moon's is 3.34.

And this is where the problem crops up. Radar and telescopic observations of Mercury implied that it had a Moon-like surface: iron silicates, with densities around 3.0 to 3.3. So Mercury was as dense as the Earth, but it looked like the Moon.

Evidently, then, Mercury is chemically differentiated, with a dense, iron-rich core underneath the layer of

A densely cratered region on Mercury, photographed by Mariner 10 *on its second encounter with the planet, in September 1974.*

The left side of this photograph is taken up by part of the huge basin named Caloris, the largest structural feature on the planet Mercury. Some 1300 kilometers (800 miles) across, this ring basin is bounded by mountains two kilometers high. The basin floor is filled with cracks and ridges. Astronomers believe that Caloris was created by the impact of a meteor at least ten kilometers in diameter.

lighter iron silicates. Scientists estimate that the top layer is some 500-600 kilometers deep, with the iron-rich core comprising the other 75-80 percent of the planet*. This core shrank slightly during the crater-formation period, causing the surface to wrinkle into scarps. There is evidence that a similar process took place on Earth.

There are two schools of thought as to what process was responsible for the core's shrinking. One believes it was caused by the final cooling of the core. The other believes that the surface must have been too hot (thus still elastic)

---

*Earth's iron core extends out to only some 55 percent of the planet's radius, but the iron is under great pressure and is therefore more dense than that in Mercury's core.

at that time for the scarps to have formed. On a cooled Mercury, the shrinking could have been caused by the core, under pressure, settling into a more compact crystalline structure.

An interesting observation about Mercury, that was not entirely unexpected, is that the lava-filled basins seem to be concenetrated on one side of the planet. The same phenomenon has been observed on Mars and the Moon—and even Earth: if you look down on the globe from directly over western Russia, you see mostly land mass; if you look down from directly above the South Pacific, you see practically nothing but water*. No satisfactory explanation has yet been offered for this asymmetry.

On the Moon, the lava-filled basins are concentrated on the side that faces the Earth. Mercury's basins are gathered around the side that faces the Sun at aphelion. This doesn't explain anything, of course, since neither Earth nor Mars is a slave to tidal friction, and they both exhibit the same sort of asymmetry.

The second pass of *Mariner 10* by Mercury photographed an odd-looking region to which astronomers have given the name "the weird terrain." The surface there is very rough and jumbled; crater rims are broken down and show deep downhill gouges that seem to have been caused by great landslides. It seems obvious that this was the scene of some localized catastrophe.

A clue to the cause of the destruction may be found in the fact that the weird terrain is exactly on the opposite side of Mercury from the largest impact basin we have yet observed (Caloris Basin, see below). Similar terrain exists on the Moon, opposite the impact basins Mare Orientale and Mare Imbrium. It may be that the collision that resulted in such a large feature formed great shock waves that travelled around the planet in all directions and came together on the opposite side with enough force to shake mountains apart.

The first three Mercurian features to be given names were Caloris Basin and two craters, Kuiper and Hun kal.

Caloris Basin is the largest feature yet seen on Mercury; a circular basin about 1300 kilometers in diameter, larger

---

*According to the *Guinness Book of World Records* (Bantam, 1973), there is a spot in the South Pacific at 48°30′ S., 125°30′ W. that is in the center of a circle of islandless water, covering an area of 8,675,000 square miles (larger than the U.S.S.R.).

The bright crater on the rim of the older crater in the center of this photograph has been named "Kuiper," after Prof. Gerard P. Kuiper, a pioneer in planetary astronomy and a member of the Mariner 10 *scientific team. (Kuiper died 23 December 1973 while the spacecraft was on its way to Venus and Mercury.) The crater is 41 kilometers (25 miles) in diameter.*

than Texas. Hotter than Texas, too, since the Sun hovers directly overhead at aphelion, driving temperatures up to 560°F., higher than the melting point of tin.

The crater Kuiper is the brightest spot yet observed on Mercury, with an albedo of 0.23. It was named after Gerard P. Kuiper, a pioneer in planetary astronomy, who was involved with the *Mariner 10* probe but died before it reached Mercury.

Hun kal is a small crater, less than a mile in diameter, that is used as a reference point in the *Mariner 10* coordinate system. It lies one degree south of the Mercurian equator, and is exactly on the 20° (longitude) meridian. Anyone who thinks scientists lack a sense of whimsy should try to figure out what "Hun kal" means: it is the word for the number 20 in the Mayan language. The Mayans used a base-20 number system and worshipped the Sun.

There are years of work ahead, analyzing the data from *Mariner 10*. And more data may be forthcoming, if the spacecraft continues to function, since its orbit around the Sun is such that it passes by Mercury every two Mercurian years.

But it's not too soon to start thinking about what the next stage of Mercurian exploration will be, or indeed whether there will be another stage.

If Mercurian exploration were to follow the pattern established by lunar exploration, *Mariner 10* would be followed by a soft-landing probe, and then by human exploration. And as far as the advancement of scientific knowledge is concerned, it would be well worth it*.

Of course, men didn't rush to the Moon just out of scientific curiosity. If it hadn't been for political motivations, we'd still be sitting around talking about it, and trying to raise the money. And there will certainly never be any political pressure over putting a man on Mercury—which is a good thing. At the present state of the art, such a project would only be a stunt, dangerous and awesomely expensive.

Still, it doesn't seem unlikely that one day there will be men and women on Mercury. In fact, it would be a logical planet on which to establish a permanent base. Such a research station would have a dual function, extending our knowledge of the Sun as well as of Mercury.

Wouldn't it be more logical, though, for a solar laboratory to simply be in orbit? It could go closer to the Sun than Mercury does, and intuitively it sounds like a less expensive proposition. But there are several objections to it that a Mercury base would solve.

The most important is safety. The impressive success of the Skylab project might make us blasé about this, since the crew lived aboard such a laboratory for months with no serious ill effects. But the Earth's magnetic field protected them from the Solar System's most deadly danger: the blistering radiation from solar flares, which occur at sporadic, unpredictable intervals. This radiation is intense enough to kill an astronaut even at Earth's distance from the Sun. On Mercury it would be like sunbathing in the light of a hydrogen bomb.

---

*It's interesting to note that the data from the lunar Surveyor flights, which successfully soft-landed on the Moon, were still being evaluated when they suddenly became obsolete: superseded by the superior data that the Apollo astronauts brought back.

The southwest quadrant of Mercury, photographed 29 March 1974 when the Mariner 10 *spacecraft was 198,000 kilometers (122,000 miles) from the planet's surface.*

Another problem is that so remote a research installation ought to be self-sufficient, at least to the extent that it need not be continually resupplied with food, air and water. Nothing can be recycled with 100 percent efficiency; eventually, raw materials have to be put into the system. And the only raw materials available to the orbiting station would be a surfeit of sunlight and vacuum, with an occasional ion from the solar wind.

There are also logistic problems associated with working in zero-gravity, although it has its compensations (one Skylab scientist reported that it took him several days, once back on Earth, to get used to the fact that things would fall to the ground if you let go of them). The station could be designed to spin, giving it a kind of artificial

Six hours before its closest approach to Mercury, Mariner 10 *took these 18 photographs at 42-second intervals, to provide a photomosaic of the planet's southern hemisphere.*

gravity through centripetal force, but this raises other problems.

To see how a permanent Mercurian base would be superior, let us try to imagine how it would be constructed, and what it would look like.

The base itself would be underground with a few meters of soil insulating it against the heat and cold and protecting it from deadly radiation. Power for the base would come from the Sun, of course, but not directly because of the 88-day night. There would be an energy collector in orbit, in the form of an array of solar cells attached to a laser. Every time it passed over the base, the laser would beam down its accumulated energy in concentrated form.

An underground farm would provide food and, through photosynthesis, oxygen for the research team stationed at the base. Water would be their most precious resource, and would be recycled with maximum efficiency.

Much of the solar research might be automated with various monitoring instruments sitting on top of the base, but our experience on the Moon shows the superiority of having a human in the system when it comes to exploring a planet's surface.

Investigation of the Mercurian surface and subsurface will probably be done with the aid of a vehicle similar to the lunar rover. Unless it were heavily armored, though, little exploration could be done during the long day, because of the possibility of a sudden solar flare.

The people at the base would probably have rather long tours of duty, on the order of years, because of the expense involved in sending a manned spacecraft to Mercury. Besides being highly trained in their scientific or engineering specialties, the staff of the Mercurian base would need a special kind of courage. Unlike the men who first sailed around the world, or explored the polar wastes, they would be in steady communication with home. Should an emergency arise, they could have the most expert advice available, within minutes...

But only advice. There would be no rescue party to follow your tracks through the snow; no hope of pointing your lifeboat toward some uncharted island. Just the irony of perishing alone in front of billions of sympathetic spectators.

HAL CLEMENT   (Harry Clement Stubbs) was born in Somerville, Massachusetts, in 1922. After attending public schools in Arlington and Cambridge, he graduated from Rindge Technical in 1939. He received a B.S. in astronomy from Harvard in 1943, M. Ed. from Boston University in 1947, and M.S. in chemistry from Simmons College in 1963.

He joined the Air Force during World War II, and flew 53 missions over Europe as co-pilot and pilot of B-24's, with the Eighth Air Force. Still a member of the USAF Reserve, he carries the rank of colonel.

Clement's first story was published in 1942, while he was still in college. His science fiction novels, such as *Needle* and *Mission Of Gravity,* are classics in the field. He has also written many science articles and a regular column on science books for *Horn Book* Magazine. And, after he "got tired of drooling over paintings I couldn't afford," he began to paint, mostly astronomical subjects. He has sold more than fifty paintings under the name of George Richard since 1971.

Clement is married and has three children. He teaches chemistry and astronomy at Milton Academy, in Massachusetts. He has been active in Boy Scout work and served as a member of the Milton town finance committee. He has been a judge for many years for the New York Academy of Science's Children's Book Prize. He is a member of the New England Association of Chemistry Teachers; Association of Lunar and Planetary Observers; Meteorological Society; New England Science Fiction Association; Science Fiction Writers of America; Boston Author's Club.

Jupiter: Eden With a Red Spot

Hal Clement

	JUPITER	
mean radius	equatorial: 71,350 kilometers	Polar: 67,000 kilometers
mass†	317.8	
mean density	1.33 grams/cubic centimeter	
surface gravity†	2.7	
escape velocity	61 kilometers/second	
length of day*	9 hours 50 minutes	
length of year*	11.86 years	
inclination of orbit to ecliptic	1°.3	
inclination of equator to orbit	3°.1	
mean distance from Sun	5.20 AU = 7.78 x 10^8 kilometers	
eccentricity of orbit	0.048	

satellites	diameter (km)	distance from planet (millions of kilometers)	orbital period (days)*	date of discovery
Amalthea (V)	140	0.181	0.50	1892
Io (I)	3700	0.422	1.77	1610
Europa (II)	3600	0.671	3.55	1610
Ganymede (III)	5270	1.071	7.16	1610
Callisto (IV)	5050	1.884	16.68	1610
Himalia (VI)	120	11.50	250.63	1904
Elara (VII)	40	11.73	259.65	1905
Lysithea (X)	20	11.83	260.50	1938
Leda (XIII)	14	12.4	239	1974
Ananke (XII)	20	21.0	625	1951
Carme (XI)	24	22.5	692	1938
Pasiphae (VIII)	40	23.5	737	1908
Sinope (IX)	36	23.7	755	1914

† Earth = 1
* Earth time

It is easy to see why the fourth planet was named after a god of war by natives of the third. It is not quite so easy to guess why the fifth received the name of the chief of the deities. The people who called it Jupiter knew none of the facts which would have justified the title; to the Greeks and Romans, as to the Egyptians and Babylonians before them, it was just another of the moving lights in the sky—neither the brightest nor the faintest, the fastest-moving or the slowest; it did not even have an eye-catching color.

Conceivably, the steadiness of its light may have been the deciding factor. Mercury and Venus change their brightness drastically as a result of changing distance and

Jupiter, as never seen before. Details of the planet's cloud cover that could not be seen from Earth are photographed by the Pioneer 10 *spacecraft while still 1,840,000 kilometers (1,121,000 miles) from the Giant Planet.*

138

phase. Mars is nearly twice as far from us at bad oppositions as at good ones, and near conjunction with the sun is some six times as far as at the best dates; its light varies accordingly.

Jupiter varies only about 20 percent from its average distance; and while the figure for Saturn is even smaller, the changing attitude of the latter's rings makes a difference easily noticed by the interested sky watcher.

Of course, interested sky watchers form only a small part of the population. To the average citizen Jupiter is merely one of the brighter stars, if he happens to notice it at all. Its twelve-year looping motion around the sky map does not catch attention, except from people who have been told, or have convinced themselves, that some things away from the earth merit their serious thought. Such people have been around for many, many centuries. The reasons for their interest in the sky may seem rather silly now, but the interest was nevertheless real. The motions of the planets were noted, recorded, and described long ago. The explanations offered for them, and the implications suggested to underlie them, were of course mystical by our present standards. They were gods; their positions were guiding fingers to those wise enough to be guided; they were evidence of the Almighty's handiwork.

But until about the sixteenth century, they were not worlds. There was only one World, and that had been placed in the center of the Universe for Man's convenience.

Then came Copernicus with his blasphemous attempt to demote The World to the status of a mere planet, an idea which worked both ways; and Galileo with his telescope, which showed planets to be round as all educated men by this time knew The World to be; and Kepler, and Tycho, and Isaac Newton. It seems safe to say that "rational" ideas about the planets date from Copernicus, while "real" knowledge about them started with the telescope. Certainly it is the telescope which provides our first meaningful ideas about Jupiter.

A small instrument—say, three or four inches aperture—shows the fifth planet clearly as a round object, not the mere diffraction-spiked point of a star. The disk is generally light in tone—grey or tan as the observer's mind tends to interpret it — and just detectably elliptical rather than circular. There will perhaps be a couple of darker streaks across it, parallel to the longer axis of the ellipse. As many as four starlike objects appear strung out at intervals along the projections of the same axis on either side of the disk, up to a dozen or so times the latter's width away; these are the four great moons, called the Galilean satellites after their discoverer. All four are not always visible; it is common to have at least one of them, and sometimes more, out of sight behind the planet, or in its shadow, or between us and the disk where a small telescope will not reveal them.

Sometimes even a small telescope will show other details. In the mid-seventeenth century one of the instruments of the time revealed a spot to the south of the long axis (it may or may not have been the now famous Red Spot). This object drifted across Jupiter from east to west (as the star maps had it; since we are looking from underneath, these show west at the right when north is at the top), disappeared, and reappeared a few hours later at the east side once more. It provided perhaps the first real physical knowledge of Jupiter: the globe was rotating, just as the more courageous scholars admitted the earth to be, though much more rapidly. Jupiter's "day" evidently measured about ten hours from sunrise to sunrise.

More details show through larger telescopes. The bright "zones" and dark "belts" show irregular edges and smaller spots of contrasting appearance. These are also carried around by the planet's rotation, but in addition shift in position, shape, and color on a less regular and predictable basis. It became evident fairly early in the history of telescopic astronomy that we were not looking at a solid surface. It might be a solid layer of clouds in an atmosphere, or varicolored material floating on an ocean. Interestingly, while both ideas must have occurred to astronomers about the same time, no one has ever seemed to attach much weight to the ocean interpretation. Perhaps this has been because the relative motions of some of these spots have been up to 320 kilometers an hour, and it is easier to imagine winds of that speed than ocean currents.

Trying to get a more precise measure of Jupiter's rotation period by tracking some of the smaller features, however, tends to give inconsistent results. Different latitudes seem not to travel at the same speed. There may be a real, solid surface under those clouds, but if so its rate of rotation is open to some doubt.

Features more than 45 degrees from the equator do

The planet Jupiter, photographed by the Pioneer 10 *spacecraft on 1 December 1973 from a distance of 2,500,000 kilometers (1,550,000 miles). The famous Red Spot is easily seen at the left, near the day-night terminator. The dark spot near the center of the planet's disc is the shadow of Io, one of Jupiter's thirteen moons.*

seem to show a day length of nine hours, 55 minutes, and about 40 seconds, which may represent the "real" day. However, marks at the equator get around about five minutes faster, a narrow region at about 20 degrees south latitude is between three and four minutes slower, and at 25 degrees north the period is a minute or so faster even than at the equator.

Even the latitude difference is not the whole story. The Red Spot, centered about 22 degrees south, does not maintain the same rotation period at all times. Over the years it has changed enough to imply a drift entirely around the planet either eastward, westward, or both, depending on the period one chooses for a background. The Red Spot cannot possibly be either a solid feature itself or the atmospheric clue to any fixed object such as a volcano below. Other spots, at widely differing latitudes, behave in comparable fashion. The simple fact must be that Jupiter is not solid either at the visible "surface" or for a very considerable distance below this level. Both the zones and the bands are fluid features whose overall structures are presumably influenced by the planet's rapid rotation (as our trade winds are by Earth's slower spin), but having details on a much smaller scale which are much harder to understand and explain. Certainly there must be complex vertical currents as well as the horizontal winds. Our own atmosphere has such motions, marked by fair-weather cumulus clouds and thunderstorms—or should we be letting our judgment be influenced by such everyday knowledge and the resultant prejudice? Maybe we should get back to observational facts and suspend opinion for a while.

If the telescope is large enough, genuine colors can be seen in the Jovian clouds. They were doubted for a long time. As recently as 1960 some astronomers considered them probably due to effects in our own atmosphere. However, the refraction phenomena which could cause such deceptions can be offset by suitable optical equipment. Large telescopes so modified, such as the 61-inch reflector of the University of Arizona's Lunar and Planetary Laboratory, have been used with color film to place the question beyond the reach of argument.

There are browns, reddish-browns, yellowish browns; yellows; blues and greens; white and greys if it is proper to call these colors. The Red Spot varies with time from bright, unarguable brick-red to a barely distinguishable pink. Verbal description is as difficult and as futile for Jupiter as it is for New England fall foliage. Even the sweeping, subjective, rather meaningless word so often used for the latter applies here; Jupiter is a very beautiful object. I don't go wholeheartedly with the more teleological philosophies, but it would be a real pity if no one were there to appreciate it. A little later I will offer my arguments for thinking it likely that someone is.

The idea that the other planets might be habitable worlds is of course not new, even in its more material and less mystical forms. At first it was presumably confined to objects of evident size such as the Moon, but the telescope quickly widened the scope of the notion. Perhaps inevitably, early thinkers tended to take for granted that a world is just another world. Many imaginative (?) writers had their characters walking around on Mars or Saturn dressed in ordinary clothes, breathing the local air, and, embarrassingly often, talking the story-teller's native language with completely human natives. This is not to say that there were no oddities and marvels on the other planets—there were plenty of both; but essentially the Solar System was for a while simply an extension of the Unknown Lands which have always been part of the storyteller's standard equipment.

To be fair, it was quite a while after the invention of the telescope before the scientists, much less the story-tellers, had any basis for a rational guess at what any planet was really like. Things seen through the telescope were interpreted in normal human fashion (I do not mean this pejoratively) on the basis of what the observer thought various parts of the earth would look like from telescopic distance. Hence the "seas" on the Moon and Mars. When the object was wholly unfamiliar and the telescope inadequate to provide details, the observer may have done his best, but interpretation—on the basis of existing ideas—did tend to creep in. I doubt that any human being has the ability to keep it out under such circumstances. Anyway, look at Galileo's early attempts to draw Saturn—or compare Percival Lowell's drawings of Mars with the *Mariner 9* photographs.

But as the seventeenth century gave way to the eighteenth and then to the nineteenth, our general understanding of physical laws and our development of mathematical tools both progressed. The relative scale of the Solar System was learned early; Kepler, at about the same time that Galileo was starting to use the telescope, knew that Jupiter was about five times as far from the sun as the earth is, though he could not have put either distance into miles very confidently. By modern logic, Jupiter should be a good deal colder than Earth. That sort of logic, however, demands a pre-existing opinion about the amount to which the earth's temperature is due to the sun, which in turn demands a fairly good quantitative understanding of the laws of radiation and of thermodynamics.

As a matter of historical fact, it was believed even in the mid-nineteenth century that the earth's temperature was due largely if not entirely to its internal heat. Jules Verne, certainly an educated and informed citizen of his time, very evidently had this impression; and whatever mistakes he may have made about weightlessness on moon vehicles and the effects of being shot from a cannon, he kept in fairly close touch with the scientific opinion of his day.

The theory of planetary origin at the time assumed that the planets had been spun off from a shrinking, rapidly rotating sun, and the outer planets, though formed earlier, would have cooled more slowly because of their greater sizes. By Verne's time the actual scale of the Solar System was quite well known—Jupiter was not merely about five times as far from the sun as the earth; it was 778 million kilometers, on the average, from Sol. Its distance from Earth could be calculated for any time, and with that distance its size was also a matter of straightforward arithmetic—11.2 times that of Earth around the equator, 10.5 times through the poles. With over 1,200 times our planet's volume but little over 100 times as much area to lose heat from, Jupiter would have to cool far more slowly than tiny Earth or tinier Mars. It could easily be warm enough for comfort. It could, for that matter, be much too warm for comfort.

An idea corollary to this fact and eagerly grasped by the story-tellers—though by no means confined to them—made the four Galilean satellites the stages for adventure, with Jupiter playing the part of a sun to keep them habitably warm. After all, the Jovian clouds could be steam or volcanic exhalations, and the Red Spot the remnant of a once planet-wide incandescence. The Spot could even be a break in the clouds which elsewhere concealed a *still* planet-wide incandescence. This idea is certainly less in conflict with modern thermodynamic knowledge than one held by a famous and competent astronomer at the beginning of the nineteenth century: he opined that sunspots were breaks in a luminous envelope through which the darker and possibly habitable real surface of the sun could be seen.

(I have used the word "real" several times in this chapter, and will, regrettably, have to use it again. I am aware that like "liberal" and "ecology" it has been inflated to near meaninglessness; I will therefore try to use it only in context that will remove any ambiguity. I recognize the growing tendency to speak of different "realities" as though any product of the human imagination deserved that title. For the benefit of mystics, acid-droppers, and flying-saucer fans who may be tempted to quote me out of context, "reality" to me is the one which the physical sciences are striving to identify and describe—the one in which you are likely to be killed if you don't bother to steer your car.)

At any rate, the notion of Jupiter as an infra-sun was popular for quite a number of years; and as far as observational knowledge went, it was quite realistic. It did not obviously conflict with any of the information arriving on *visible* light. However, there are other kinds of light, which we were gradually learning to exploit.

In the early 1920's, advances in techniques for detecting and measuring longer wave "infrared" radiation made it possible to calculate an apparently realistic value for the big planet's temperature. This turned out to be about 150° Kelvin—that is, about 120° below zero Celsius, or over 180° below zero on the Fahrenheit scale still used in less scientific circumstances. It was clearly understood by the investigators, but apparently not so clearly by others, that the measures referred to some part of Jupiter's atmosphere well above the visible clouds. Even recognizing this, however, it was hard to see how anything with very much heat down below could be that cold above. The notion of Jupiter as a minor sun was quietly buried.

Like Dracula.

The hopeful story-teller could still assume that Jupiter was habitably warm somewhere below the clouds, however, and some of them continued to do so. Human heroes and heroines continued to be kidnapped by wicked Jovians, and to confront their captors face to face with neither party wearing any special protection from temperature or atmosphere. However, facts continued their effort to discipline imaginations.

The next blow to free imagination was a pincer attack mounted from one side by a device called the spectrograph, and from the other by the same sort of story-teller's wishful thinking which produced the Jovians in the first place.

The limitations as well as the powers of the spectrograph must be kept in mind in evaluating the evidence it provides, naturally. The instrument produces a record of the various wave lengths present in a beam of incoming radiation. Essentially it makes up for the fact that the human eye is a hopelessly bad frequency analyzer, compared to our sense of hearing. Anyone can tell if a single instrument in an orchestra is out of tune; a musician of even moderate competence can identify all the instruments even when they are playing together; most people can recognize a large number of individuals by voice alone.

The eye, on the other hand, can see little difference if any between the green of a field of grass and the green of a well-processed color photograph of the same field, though the actual mixtures of wave lengths reaching the organ from the two are grossly different.

Even if the eye could analyze better in its range of

sensitivity—the so-called "visible light"—it would still be of very limited use for recognizing materials on distant planets. Light is emitted and absorbed under such circumstances that elements in the gas state, with their atoms relatively far apart, do show very distinctive wavelength patterns (spectra); compounds, however, with their atoms in the clumps we call molecules, produce more complicated patterns called *bands* which are much more difficult to distinguish and identify; while solids and liquids, whose atoms are crowded together in very large numbers, lose nearly all their atomic individuality and have very broad, very vague, very unidentifiable spectral patterns.

The last difficulty is countered to some extent if a broad enough range of wave lengths can be studied, but the human eye lacks the necessary scope. Most of the solids and liquids which we claim to recognize by color are submitting other evidence not so consciously considered, such as the shape of the bottle (or the trade mark). *Context* is always there—or when it isn't, we start to feel unsure.

Furthermore, the context occurs in our own minds as well as in the real world. It includes *what we already believe*. More sadly from the scientific point of view, it may also include what we want to believe.

At least some of Jupiter's light-reflecting materials, such as whatever colors the clouds, would seem to be either solids or liquids. There seem to be no elemental gases whose spectral patterns lie in the wave lengths which penetrate our own atmosphere. There are, however, some recognizable molecules.

Jupiter's spectrum naturally includes that of the sun; we see the planet by reflected sunlight, and wave lengths missing from the latter are not restored by bouncing from a planet. Still others are removed, however; the Jovian spectrum contains band patterns of the sort we long ago learned to associate with fairly small molecules in the gas state. In the 1930's these were specifically identified as being due to ammonia (NH_3, for those who think more comfortably in chemical formulae) and methane, CH_4.

Both of these occur naturally on our own planet, as well. Both, interestingly, are produced by life forms, albeit microscopic ones. Decay of nitrogen-rich waste matter is responsible for the ammonia odor around the diaper pail so familiar to people in the early days of parenthood.

Other microbes working on the dead organic matter buried away from the air in swamps produce the "marsh gas," mostly methane, which bubbles up from such places. Still other microorganisms living symbiotically in the digestive tracts of ruminants cause even the most courteous cow to belch five or six gallons of methane-rich gas a day as her tiny helpers predigest her grass for her.

Both compounds have practical uses in our culture. Weak water solutions of ammonia are used as cleaning fluids, and the natural gas which may possibly last to the end of this century as a fuel is mostly methane.

The identification of these gases was of course one of the points of the pincer attack on an Earthlike Jupiter, mentioned a page or two ago. The other was something of a Fifth Column movement, in which the story-tellers themselves abandoned the old field and began to make active use of the ammonia, methane, and low temperature environment with which the more disciplined imaginations of the physical scientists had presented them.

Actually, a few more physical facts contributed to the new picture. As mentioned, the physical dimensions of Jupiter were well known. Its mass could be, and had been, calculated not only four independent ways from the motions of the four Galilean satellites, but from those of numerous others discovered later as telescopes grew bigger. Therefore, its average density could easily be determined. Finally, while the theory of planetary origin was quite different from that of the nineteenth century, there *was* such a theory, and it formed a mental context for painting the new picture of the Real Jupiter.

The density was low, scarcely a quarter that of Earth. The theory still assumed that the sun and planets were made from the same stock of raw material, which included a very large percentage of hydrogen. Because of Jupiter's huge mass and consequent immense gravity, little of this light material would have been lost during formation (very much unlike the earth and other inner planets).

The overall picture, therefore, was of a planet whose core was a ball of rocky material and possibly metal, similar to the earth, some thousands of miles in diameter. This was overlaid by more thousands of miles of ice, and the whole structure topped by more thousands of miles of atmosphere made of hydrogen compounds and perhaps surplus hydrogen (and, conceivably, helium, the second most abundant material in the universe). Unfortunately for observational checking, neither molecular hydrogen nor helium has spectral patterns in the wave length regions easily studied through Earth's atmosphere.

However, some additional support for the theory came from satellite motions, which indicated that Jupiter's mass was very much concentrated toward the center; the extreme equatorial bulge—the planet is 9,300 kilometers bigger through the equator than from pole to pole—which implies the same fact; the apparently negligible heat output; and of course the already mentioned spectroscopic evidence for hydrogen compounds.

It was easy to imagine the huge, cloudy protoplanet contracting in its own immense gravity field; the denser, higher-melting elements and compounds liquefying, then solidifying, and finding their way to the center of the growing world—the iron, the nickel, the oxides of silicon and aluminum which appear to have behaved the same way in the body of our own planet. The remaining oxygen combined with the hydrogen to form ice; nitrogen did the same, producing ammonia; carbon reacted to give the methane; and the rest of the hydrogen—one could only guess how much—simply stayed uncombined, to form the rest of the atmosphere. It had no way of getting away from a planet whose escape velocity, even at 150,000 kilometers from its center, is still five times our own 11 kilometers a second.

The actual figures—the diameter of the core, the thickness of ice layer and atmosphere—could be adjusted until the total mass equalled that known for the planet. More accurately, one could *try* to do this; there are major complications in the pertinent physical laws and their applications. More about these a little later.

This was the general picture which passed rather quickly into near-Gospel status with the science fiction writers of the thirties, though they of course tended to add details of their own. While more than one of the writers did make a policy of keeping up with science, or trying to, it was probably an article in the February, 1937, issue of *Astounding Stories* magazine which really established this particular Jupiter style. The piece was written by the late John W. Campbell, who himself edited the magazine for over thirty years until his death in 1971.

The article went further than the strictly scientific picture. It pointed out that Jupiter, with all these described qualities, must be an ideal place for life—well supplied with ammonia and hydrogen, and with its temperature a mild one hundred eight-odd below zero (Campbell, both in writing and conversation, liked to hold the straight face until the verbal bomb exploded). All this, as he pointed out, constituted an ideal environment for a life chemistry using ammonia as its polar solvent rather than water, and getting its energy by reducing foodstuffs with atmospheric hydrogen instead of oxidizing them as we do.

For some years after this, the Jovian landscape—in the minds of most of those who thought about it at all—was a grim and frozen surface, lit by continuous lightning (naturally it had to be lit, but no sunlight could possibly get through those clouds), swept by gales of hydrogen-methane mixture carrying rain or snow of ammonia, dotted with lakes and oceans of liquid ammonia. Usually it had more or less logically designed life forms adapted to the chemistry, temperature, and gravity. This last was usually described as the two-and-three-quarters Earth value which applies to the tops of the clouds; most authors dismissed as insignificant the depth of the atmosphere even when they paid lip service to its "thousands of miles" quoted value.

This was a mistake, not only because of the gravity. In fact, serious thought about the physics of a thousands-of-kilometers-deep atmosphere raises a question about whether Jupiter could be said really to have a surface at all.

The writers, of course, assumed that it did; there had to be something for their characters to walk on. They even tended to assume that it was not merely the ice of scientific theory, but had somehow picked up heavier materials such as silicates, and had rocks and even soil of a sort. With rocks in general being so much denser than ice, this may seem unreasonable; but there was some excuse for it.

This stemmed from some of the physical and chemical efforts to explain the cloud colors—not everyone, even forty years ago, dismissed these as illusory. One of these theories suggested that there were traces of sodium and potassium metals dissolved in the droplets of ammonia which presumably formed the principal cloud constituent. Certainly if this were true, there must be relatively heavy elements well outside the Jovian core; there could indeed be rocks and soil in or on the ice.

The color theory itself was of course based on sound chemistry; alkali metals do indeed dissolve in liquid ammonia and do impart various colors to it. However, there are some implications to the presence of these metals which do not seem to have received much publicity. First, the solutions are not stable. Even at the supposed Jovian temperatures, the dissolved metals react slowly with ammonia to form amides and free hydrogen. The colors would bleach very quickly—by geological time standards—*unless there were some continuous source* of these extremely active metals. This is not really very much more severe a condition than having to provide them in the first place.

Apparently it was taken for granted that such a source did exist; the story backgrounds sometimes included sodium explosions (sodium burns in hydrogen, incidentally). One can understand the more disciplined imaginations' feeling hesitant to come up with the obvious explanation, but it is harder to see why the science fiction enthusists passed it by.

There has to be a special explanation for the presence of *any* uncombined active element in an environment which includes things it can react with. This includes not only sodium and potassium, but items from the other side of the chemists' periodic table such as oxygen, fluorine, or chlorine. On Earth, as I would like to believe every high school graduate knows, the supply of oxygen is maintained by life forms—the photosynthesizing plants. Without these, our atmosphere would lose all but traces of its oxygen within a few decades. *If* there actually turns out to be free potassium or sodium on Jupiter, I personally will take it as a mandate to start looking for life there (though I admit other causes for the reduction are conceivable).

But from the physical chemists' point of view, the atmospheric depth is just one of the details of the more general problem of what Jupiter is really like inside. The composition of the clouds is pertinent, but probably not crucial, to the major headache of making a *detailed* calculation of what lies below them. It is far more than multiplying the densities of a rock sphere and an ice shell by their respective volumes and adding the results. *The densities are variable!*

Gases, as we learned in high school, are strongly affected by pressure; as the latter increases, the volume of the gas gets smaller. The pressure in turn depends on the weight of the overlying gas, which depends on its molecular weight and the local gravity. Just to make it all more fun, the temperature distributes its influence between pressure and volume. As we descend into Jupiter, the pressure must get higher because there is more and more gas left overhead; the gravity is getting stronger as we approach the center of the planet, but weaker as more of the total mass is above us and ceases to contribute to the pull—at first, the distance effect predominates, but sooner or later the other takes over. Gravity is zero at the planet's center. The temperature is presumably rising as we descend, but we can only guess at how fast; the rate of the pressure *change* varies with gas density and gravity, therefore with temperature and current pressure...

If your head is spinning, don't feel guilty. We started in the middle. It should be simple at least to get the conditions in the outer layers, where we can see—shouldn't it?

Well, we need a temperature at that level; we were pretty sure of that for a long time. We need the gravity value; no problem. We need the average molecular weight of the gases: well, let's see. Methane and ammonia are certainly there, with weights of sixteen and seventeen on the usual chemists' scale. That would pin down an average pretty closely even if we didn't know the percentages of the two.

Unfortunately, we have already mentioned that there is reason to suspect the presence of hydrogen, of weight two, and helium with molecular weight four. We have also admitted that neither of these gases shows in the Jovian spectrum, so there is no immediate method of estimating their quantities; and this will make a very large difference in the starting point for our arithmetic. The gas layer is going to give a lot of trouble, it seems.

But solids, we learned in high school, are not compressible; they have constant densities. This, unfortunately, is only an approximation, usable under the near-vacuum conditions in which we live regularly. Even on our own planet, the operator of a bathyscaphe or other deep-water submersible has to make allowance for density changes in the water which supports his craft. The pressure 10 kilometers down in our oceans would certainly be exceeded 200 kilometers or so below Jupiter's cloud tops—say, three percent or so of the way to the

JUPITER WEATHER

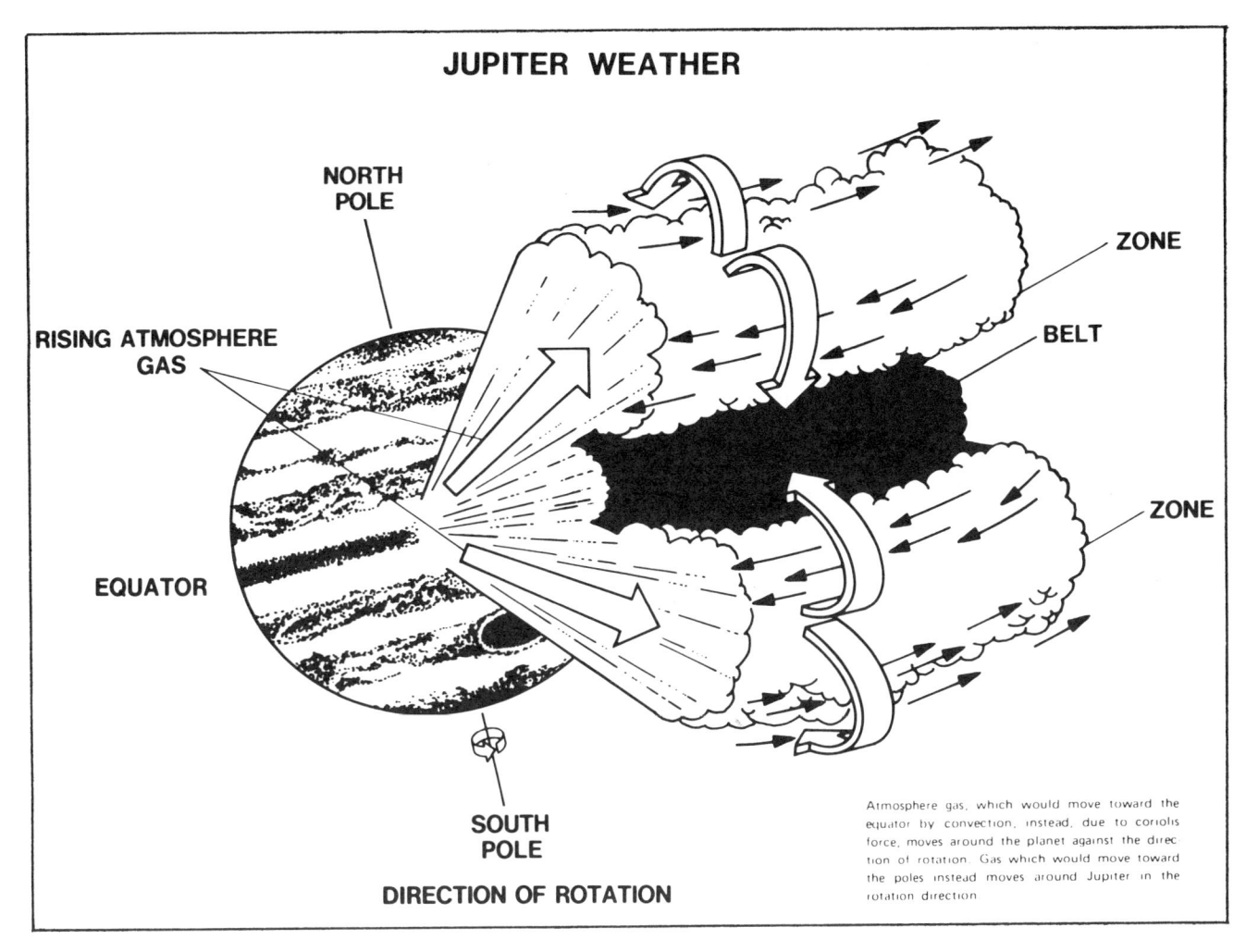

Atmosphere gas, which would move toward the equator by convection, instead, due to coriolis force, moves around the planet against the direction of rotation. Gas which would move toward the poles instead moves around Jupiter in the rotation direction.

A possible model of Jupiter's weather, based on data transmitted by Pioneer 10.

planet's center. The ice layer we mentioned, the rocky material, and even a metal core (if there is one) are all subjected to quite enough pressure to change their densities, and therefore to change the rate at which their densities change with increasing depth.

And, as if that were not trouble enough, there is the matter of phase change. At some pressure-temperature combination in Jupiter's atmosphere, ammonia changes from gas to liquid. Somewhere else, it goes from liquid to solid. Somewhere else again, methane does the same. Under high enough pressure, even hydrogen becomes a metallic solid. Ice isn't just ice; it has half a dozen or more different crystal forms depending on the local temperature and pressure. So do many, if not all, minerals. The infinity of possible *mixtures* of all these can also, in many cases, show sharp phase changes.

In our own atmosphere just one substance, water,

153

commonly changes phase—to and from solid, liquid, and gas; and that one substance is enough to complicate weather prediction to the point where even high speed computers can't work fast enough to stay ahead of real-time.

To sum it all up mathematically, there are equations which describe more or less adequately the sorts of thing which *can* happen inside Jupiter, but they involve large numbers of terms; and any high school algebra student knows what has to be done when there are more unknowns than there are equations.

I am not suggesting that a calculation of the Real Jupiter is impossible or even beyond present day mathematics, but I do say that we really need more equations—that is, more *quantitative* observations of whatever Jovian conditions we can possibly manage to observe. The fact that we don't have laboratory data for all possible materials under Jovian conditions is probably less inconvenient than the fact that we don't know just what substances we should be allowing for. The old rock-ice-atmosphere model was just a rough guess.

Fundamentally, there are two ways to increase our supply of quantitative observations. One is to increase the *scope* of our coverage, either by observing in more parts of the spectrum than we now cover or by considering aspects of the radiation such as its polarization which were not previously considered or recorded adequately. The other is to improve *resolving power*, the ability to distinguish two closely spaced points or two very slightly different wave lengths or, essentially, two very near-together anythings.

Our original temperature measures of the planets resulted from extending the scope of our measurements into the infrared part of the spectrum. This process can be carried only a short way, as long as we observe from Earth's surface; our atmosphere blocks much of the infrared part of the spectrum by the absorption bands of carbon dioxide and ozone. Increasing our ability to *resolve* wave lengths in the accessible region, however, can and has improved those temperature determinations. We were able to "penetrate" farther into Jupiter's atmosphere, and evidence for temperatures above the melting point of ice and even, in small areas, above human body temperature has been secured. In fact, the total energy radiated from the planet has been far more reliably determined in the last few years, and leads to the startling fact that Jupiter is emitting more than twice as much energy as it is receiving from the sun!

The ultra-cold planet idea has to be dropped, just as the infra-sun one had to be a few decades ago. It is a *very* feeble sun, however; whatever its inside temperature, it is still not radiating enough to influence the temperature of even its closest satellite.

The really meaningful implication of the new, hotter Jupiter is that the sun has little if anything to do with the cloud motions we have been watching for the last few centuries. Jupiter's weather, unlike those of Earth and Mars, is *internally* powered. The drive is from below, which suggests to any meteorologist that thunderstorm-like convection currents should play a very large part in the "air" circulation. If we could only see more details...

The angular resolution which this demands is the sort of thing the human eye is pretty good at, but it is working at the serious disadvantage of a 620-million-kilometer minimum distance. Worse, the eye is looking through Earth's atmosphere, which not only restricts the wave lengths we can use but also changes the direction of the rays which do get through. If the changes were constant or even consistent this would not be too bad, but they are neither. Every time a ray of light travels from a mass of air at one temperature to a mass at a different one, its direction is slightly changed. The masses do not have constant shapes or boundaries, as anyone looking across a bonfire or a pavement on a hot day is aware. The constant shimmering, trembling effect is magnified by any instrument such as a telescope which is designed to improve the eye's resolving power. A 1500-kilometer-wide thunderstorm on Jupiter would be a barely discernible speck through any Earth mounted telescope. It would seem that to make any really good observations of the planet we must not only get out of our own atmosphere but cross a good fraction of the hundreds of millions of kilometers which separate us.

This, however, is not entirely true. It might almost be said that confinement on Earth under its trembly atmosphere has been good for astronomers. It has forced them to learn the art of squeezing every bit of information still contained in the radiation which does reach them; they are photon-misers, skilled and experienced treasure-hunters among the quanta. Sometimes, too, they have a little luck. It must be admitted that luck helped a bit in solving the molecular weight problem of Jupiter's atmosphere, though luck alone would not have been enough.

Twice in the last couple of dozen years, Jupiter has passed between us and a fairly bright star. Each time, we knew that this was going to happen, and were waiting for the event with radiation-analyzing equipment ready. The rays from the star naturally shone through increasing depths of Jupiter's upper atmosphere, and were bent and scattered by it just as they are by our own. The light bending and wave length spreading depended on just the factors mentioned earlier as being needed to start the mathematical drill toward the planet's center—gravity, temperature, and average molecular weight of the gases. We knew the gravity, had a fair idea of the temperature, and could therefore calculate with some confidence the final quantity.

The answer, both times, was close to *three*. In other words, the gas mixture was mostly hydrogen and helium. The ammonia and methane which dominated the visible spectrum were relatively minor constituents, more or less analogous to the carbon dioxide and water vapor in our own atmosphere.

This, of course, was down to the cloud tops, no lower. The starlight was cut off just as the information it carried was getting interesting. Still, we could put the right sized bit in the mathematical drill; luck had done its bit, helped by preparedness.

Other information came more from engineering skill than from luck. The visible and almost-visible radiation is not all that gets through our atmosphere; there is also the so-called "radio window." (Transmission is not the simple on-and-off matter that this phrase suggests; the transparency of our gas envelope varies practically continuously as we explore along the electromagnetic spectrum. Details, however, are rather out of place here.) We have learned to exploit this part of the spectrum, especially since World War II, when the development of radar contributed greatly to our skill in this art. Radio waves, we now know, are reaching us from many sources beyond the earth; Jupiter is only one of them. It is, however, a very strong one, and a good deal of information has been gleaned about the planet with the radio telescopes.

From the nature of the radio emission detected it was

deduced that Jupiter has a vast magnetic field. This field apparently traps electrically charged particles from the sun, just as our own does, and forms Van Allen belts not wholly unlike those of the earth.

The radio emission varies as Jupiter rotates, and may be giving a "real" value of the rotation period. It also changes with the motion of the innermost large moon, Io, which was a little more surprising.

Exactly what causes all these radio waves is far from certain, but it seems reasonable to guess that much can be attributed to electrical activity not greatly different from that of Earthly thunderstorms. This at least fits in with the guess we made earlier about convection currents driven by the internal heat; but two points on a graph don't prove much about the shape of the line, and once again we are basing ideas on what we think we know already.

Regardless of the specific source, however, any such idea can be dignified with the name of *hypothesis* and

A conceptual view of Jupiter's magnetosphere, which is generated by the interaction of the planet's magnetic field and the solar wind. Although the Jovian magnetosphere is similar in nature to Earth's, it is many times stronger.

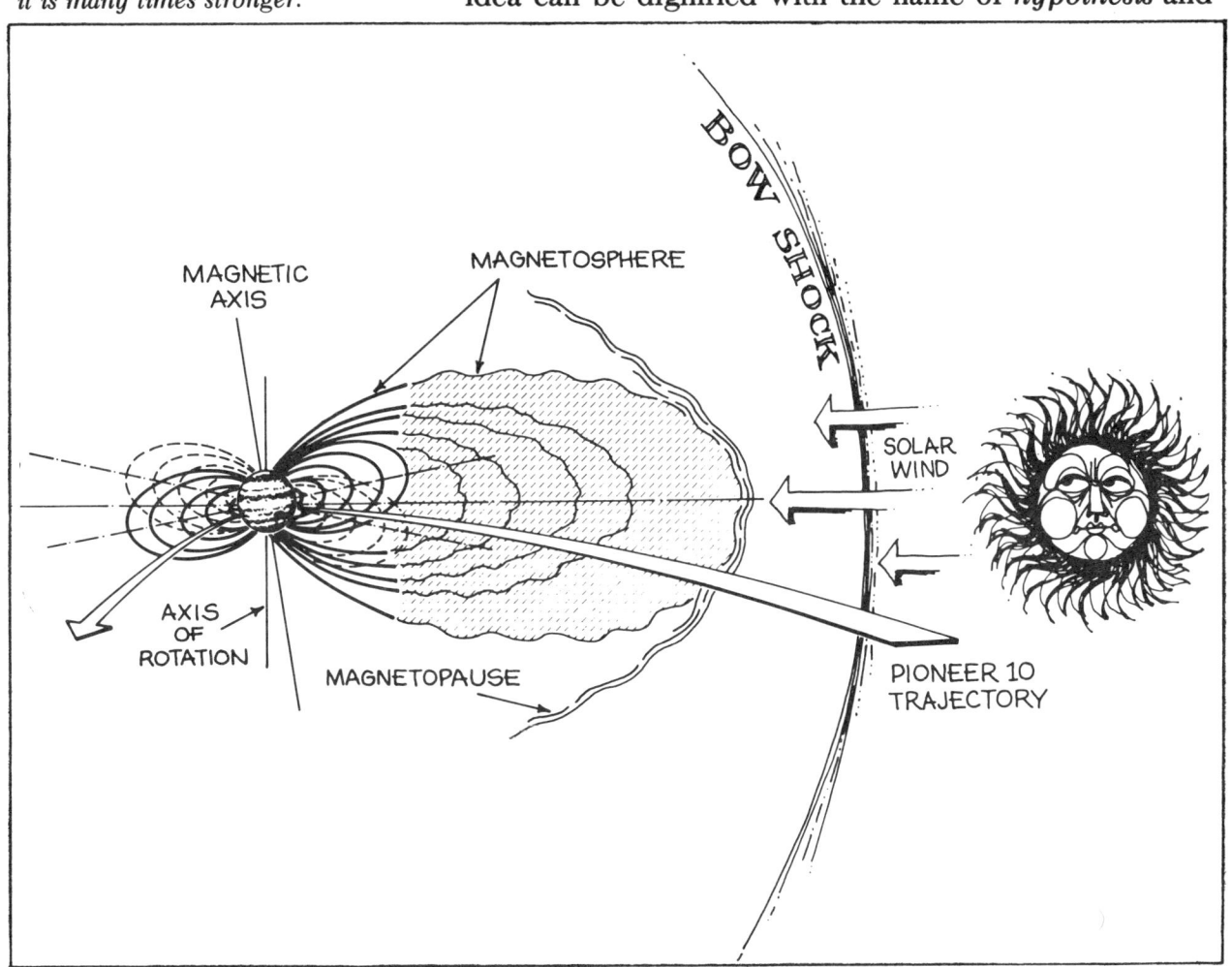

play a respectable part in the extension of human knowledge. The idea, good or bad, provides a basis for designing experiments or planning observations which may prove or disprove it (or, more likely, merely suggest whether or not it is worth following up). In the present case, the notion about convection currents suggests that it would be worth trying to measure Jovian cloud *heights*. This will still demand comparison with Earth's cloud types, if we succeed in the measurement; but we do, after all, have a pretty good idea of the physical processes which form the latter, and however imperfect analogies may be, they do have their uses. But how can we measure the cloud heights on Jupiter?

The shadow-measuring technique used to determine the heights of the moon's mountains is hopeless here, partly because of Jupiter's distance (resolving power) and partly because it is necessary to have side lighting to use the geometry effectively. Earth is never more than twelve degrees from the sun, as seen from Jupiter, so shadows are always pointing almost directly away from us and wouldn't lend themselves to measurement even if we could see them at all.

However, there are other methods than pure geometry for judging distances. Here on Earth, far-away mountains and other landscape features seem hazy and bluish in color—what artists call *aerial perspective*. It is due, of course, to lack of perfect transparency in the air, and the intrusion of extra light into a long visual path by scattering particles such as dust, water droplets, and even density variations caused by random motion of the gas molecules.

A combination of this principle with improved wavelength resolving powers did provide a fairly good method for measuring Jovian cloud height. By filtering out all the light *except* for wave lengths absorbed by methane, and using these to take photographs, we could get pictures in which the highest clouds showed their details most clearly, while those deeper in the atmosphere were correspondingly less visible. The Lunar and Planetary Laboratory has carried on a regular program of this sort for several years. It has also carried on a careful search for a wide variety of what seemed the most likely chemicals, mostly low-molecular-weight hydrocarbons and compounds of sulfur, nitrogen, and oxygen with hydrogen and with each other. Unfortunately, few if any of these substances lend themselves to spectroscopic detection in the wave length ranges which get through our atmosphere; and most of our opinions about their presence are based more on inference. In other words, they are reasonable (we think) but may not be right.

There is, as a result of all this, another consistent picture of the "real" Jupiter, more detailed than that of 40 years ago.

The atmosphere is mostly hydrogen, heavily laced with helium. It is extremely turbulent, as a result of convection currents driven by the planet's internal heat. At various heights it contains methane, ammonia, and even water in

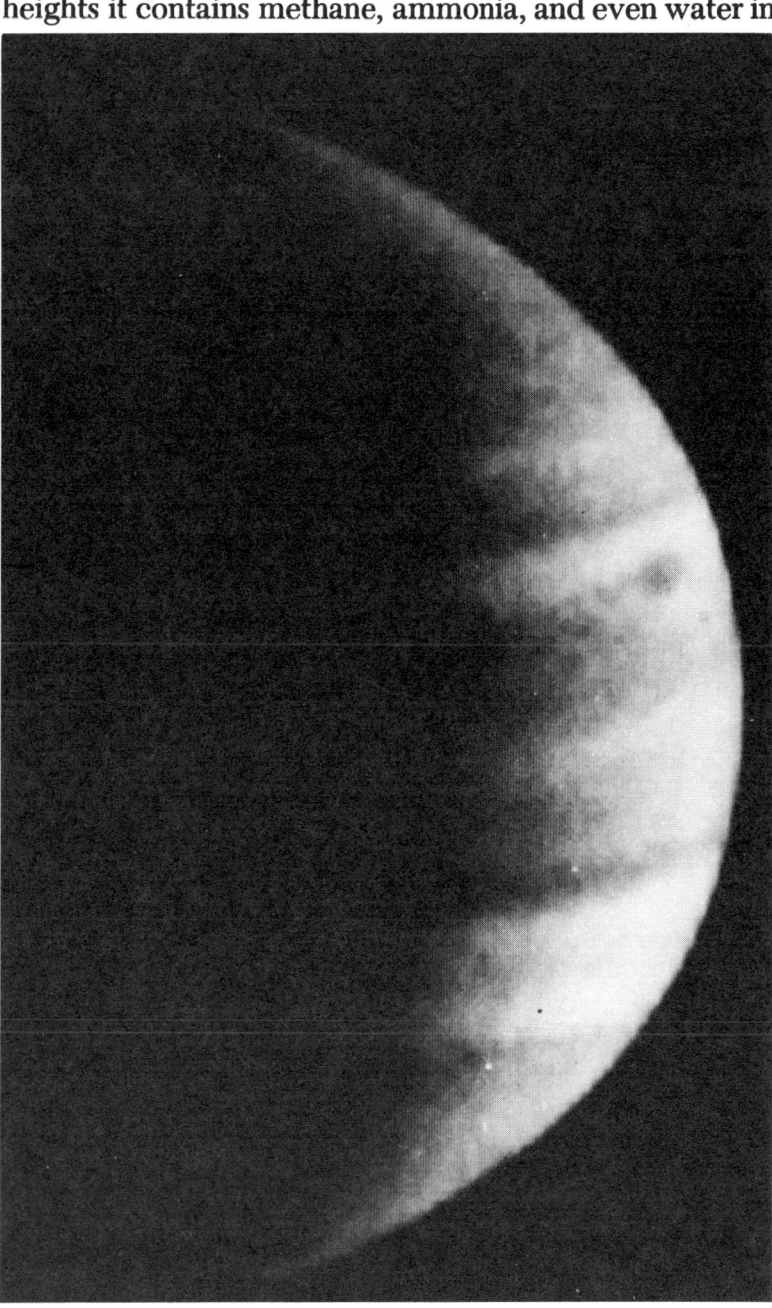

Crescent Jupiter, a view that can never be seen from Earth. Since Jupiter is so much farther away from the sun than our planet is, we always see Jupiter in its "full" phase. Pioneer 10 made this picture while 2,487,000 kilometers (1,545,000 miles) from Jupiter. The light band crossing the upper half of the crescent is the North Tropical Zone. The gray spot near its southern edge is the Little Red Spot.

gas form. High above the visible clouds, some of the ammonia is in the form of tiny, frozen crystals, forming a general haze which contributes to the aerial perspective mentioned above and helps with the determination of cloud heights. At lower levels, higher concentrations of the ammonia crystals form white or nearly white cirrus-type clouds. Lower still, hydrates of ammonia and even water ice do the same.

Hydrogen sulfide is assumed to be present. At the very highest levels it is broken down by ultraviolet light from the sun, providing free sulfur at varying shades of white and yellow according to the temperature and the sizes of the particles.

The sulfur, as it settles toward the deeper and warmer

Artist's imaginative view of Pioneer 10 *flying over the Red Spot.*

levels, combines with hydrogen again and also with ammonia; the resulting sulfides can join their tiny molecules, again under the influence of ultraviolet light, to form hydrogen and ammonium *poly*sulfides—longer, heavier molecules which can explain the browns, yellows, and oranges of some of the clouds.

The Red Spot seems to be a continuous family of convection cells—essentially, a huge cluster of thunderstorms tending to decay at its western end and develop at the eastern, so that the spot itself gradually advances around the world at a different apparent speed from the true rotation. The spot seems to be four or five kilometers higher, at least when at its most vivid, than the highest cirrus clouds of the belt in which it rides (though both height and, as has been mentioned, color intensity of this area are subject to change). Other, much smaller spots, usually white, behave in similar fashion; the tops we see are supposed to be the spread-out cirrus area, several times the diameter of the rising gas columns themselves, just as occurs in our own atmosphere.

The presence of such currents would keep the lower atmosphere from getting too much hotter than the layers above, at least for any great length of time. When such a hot spot did develop, the cooler gases above would eventually give way and permit a practically explosive updraft to form—tornado weather, one is tempted to call it. At least, this provides some guidance for our mathematical drill; it is not too hard to calculate, from the known behavior of gases, the rate of temperature change which cannot be exceeded greatly or for long. The terrific upward currents could send shock waves up into the near-vacuum layers of Jupiter's ionosphere to produce some of the radio bursts we receive.

Deeper down, the atmosphere is thought to be much less violent, with the rate of temperature change much less (in effect, the upper layers may be thought of as acting as a sort of blanket). Under these conditions, the pressure change can be calculated with some confidence, and such calculations suggest that the hydrogen may shift to a strange, low-density, metallic liquid only a few hundred kilometers down; a substance that would look like mercury if we could see it, and would float on water if liquid water were possible there. It seems unlikely, however, that there would be a clear-cut surface for science fiction characters to walk around on.

Since the "weather" seems, in this picture, to be confined to the top 180 kilometers or so where the atmosphere is really violent, it may well be that not even lightning illuminates this hard-to-imagine region where liquid and solid may not have distinct meanings.

All of this picture has been painted with the poor resolution obtainable from the Earth. Presumably, a close look at Jupiter would do much either to support or refute it. Unfortunately, our close looks have so far been very limited. Two probes, *Pioneer 10* and *Pioneer 11*, have so far approached the giant planet.

It is tempting to suggest that the first probe made startling changes in the picture described in the preceding page or two. Unfortunately, most of its work had little to do with the *picture* in any purely visual sense—though photographs were indeed sent back. It must be remembered that there was a very sharp weight limit on the equipment which could be sent with the *Pioneers*. Both

Layout of the Pioneer 10 *and* 11 *spacecraft. NASA calls each spacecraft by letter, rather than number, until the craft has been safely inserted in space. Thus Pioneer F became Pioneer 10 once it was sent on its way to Jupiter. Pioneer G (11) will fly past Saturn in 1979.*

PIONEER F/G SPACECRAFT

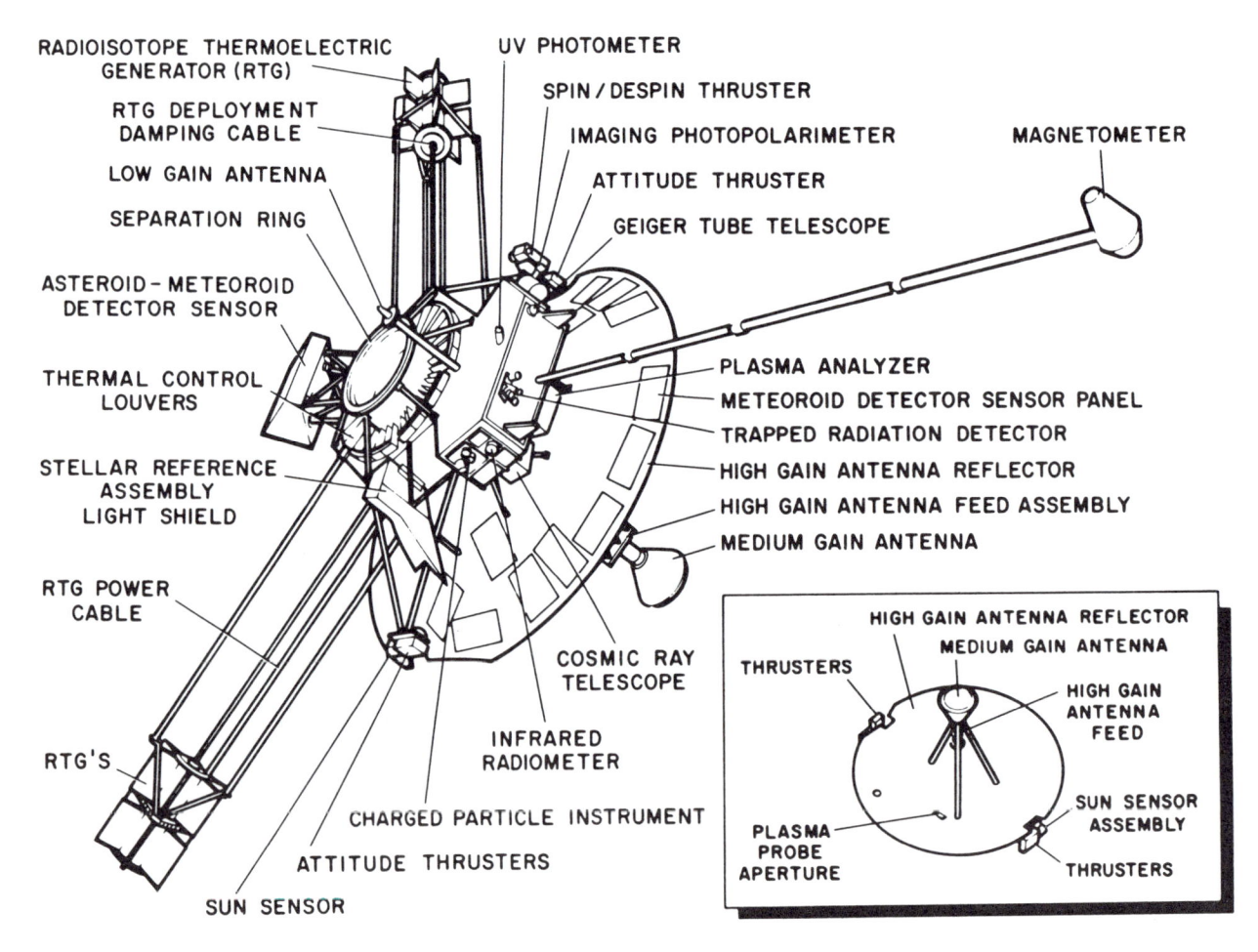

162

PIONEER F/G ENCOUNTER WITH JUPITER

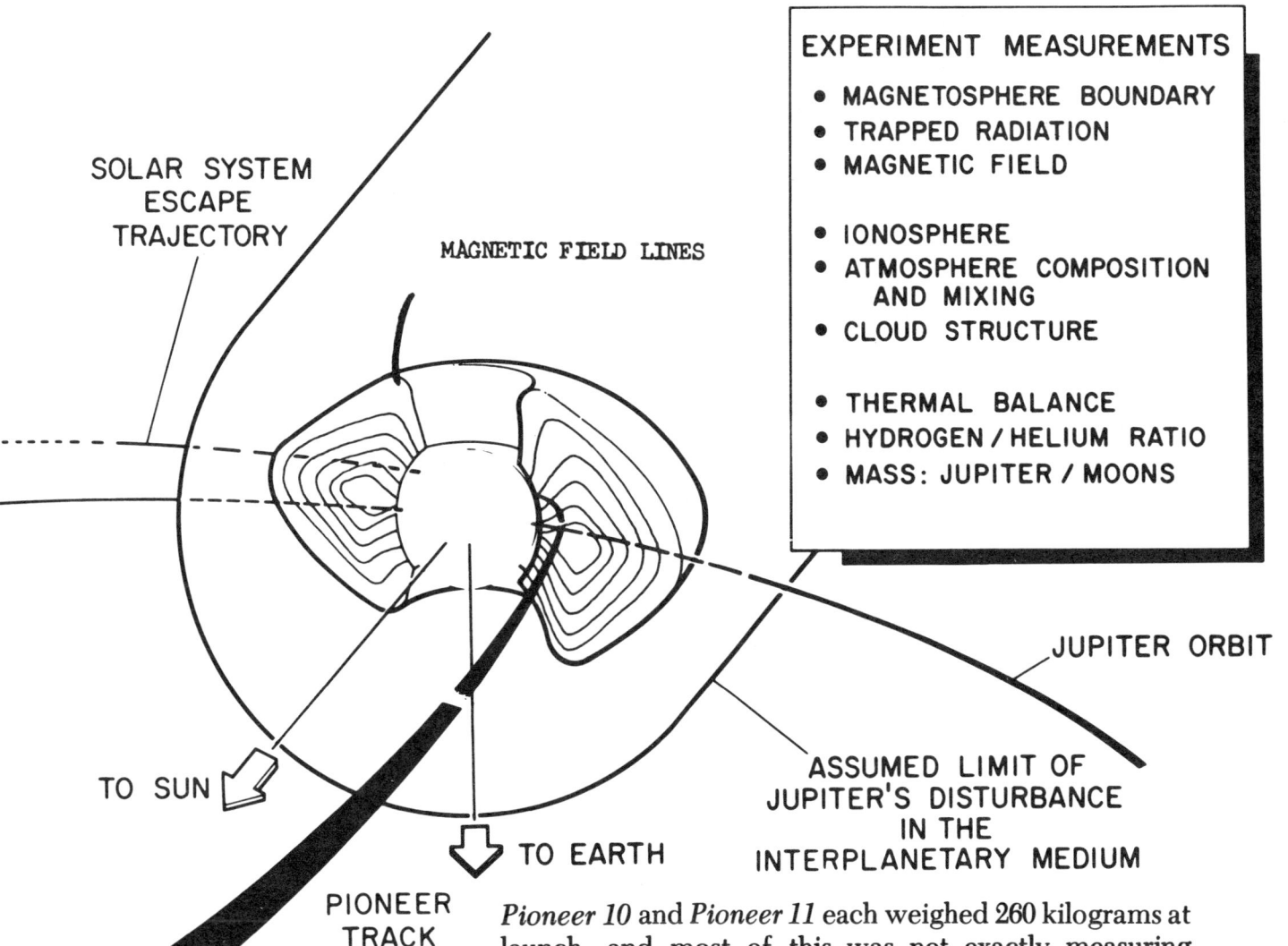

EXPERIMENT MEASUREMENTS
- MAGNETOSPHERE BOUNDARY
- TRAPPED RADIATION
- MAGNETIC FIELD

- IONOSPHERE
- ATMOSPHERE COMPOSITION AND MIXING
- CLOUD STRUCTURE

- THERMAL BALANCE
- HYDROGEN / HELIUM RATIO
- MASS: JUPITER / MOONS

Sketch of the Pioneer *trajectory past Jupiter, showing how the spacecraft encountered the Jovian magnetosphere.*

Pioneer 10 and *Pioneer 11* each weighed 260 kilograms at launch, and most of this was not exactly measuring equipment. For one thing, the solar panels used on missions closer to the sun would have delivered only 4 percent as much power at Jupiter; they had to be replaced by a pair of 60-watt nuclear generators in each of the spacecraft. There had to be communication equipment, including rather impressive antennas to send the information more than three-quarters of a billion kilometers back to Earth. There had to be guidance equipment and "steering" rockets so that cameras and other gear could be directed properly, and the orbit of the craft itself changed if necessary. Actually, the scientific experimental devices weighed only 29½ kilograms in *Pioneer 10* and 30 in its successor.

Furthermore, much of this equipment was not what a science-fiction enthusiast would have selected. Nothing was intended to search for life. Much was concerned with engineering problems; what aspects of the Jovian and near-Jovian environment would affect the spacecraft itself, and require changes in the design of future craft? There were devices to study Jupiter's magnetic field and the high energy particles which we suspected must be trapped in it—would this radiation interfere with other sensing devices, or with the guidance or transmission equipment of the space craft? There were detectors for meteoroids and asteroids—the trip lay through the asteroid belt, where nothing man-made had yet travelled. (We were quite certain that the asteroid belt was not the deadly menace to space craft which had been pictured in the stories of the '30s, when the crews of vessels penetrating the region stood double watches to avoid collision; but numerical data were still very much needed.) There was a plasma detector, and photometers for both infrared and ultraviolet light, and, admittedly, there was a "camera"—a so-called imaging photopolarimeter. The equipment of *Pioneer 11* was nearly the same.

Details of Pioneer 10's *flight through the most intense portion of Jupiter's magnetosphere. Radiation intensities from trapped particles in Jovian Van Allen Belts became so high that* Pioneer's *electronic equipment was in danger of being damaged.*

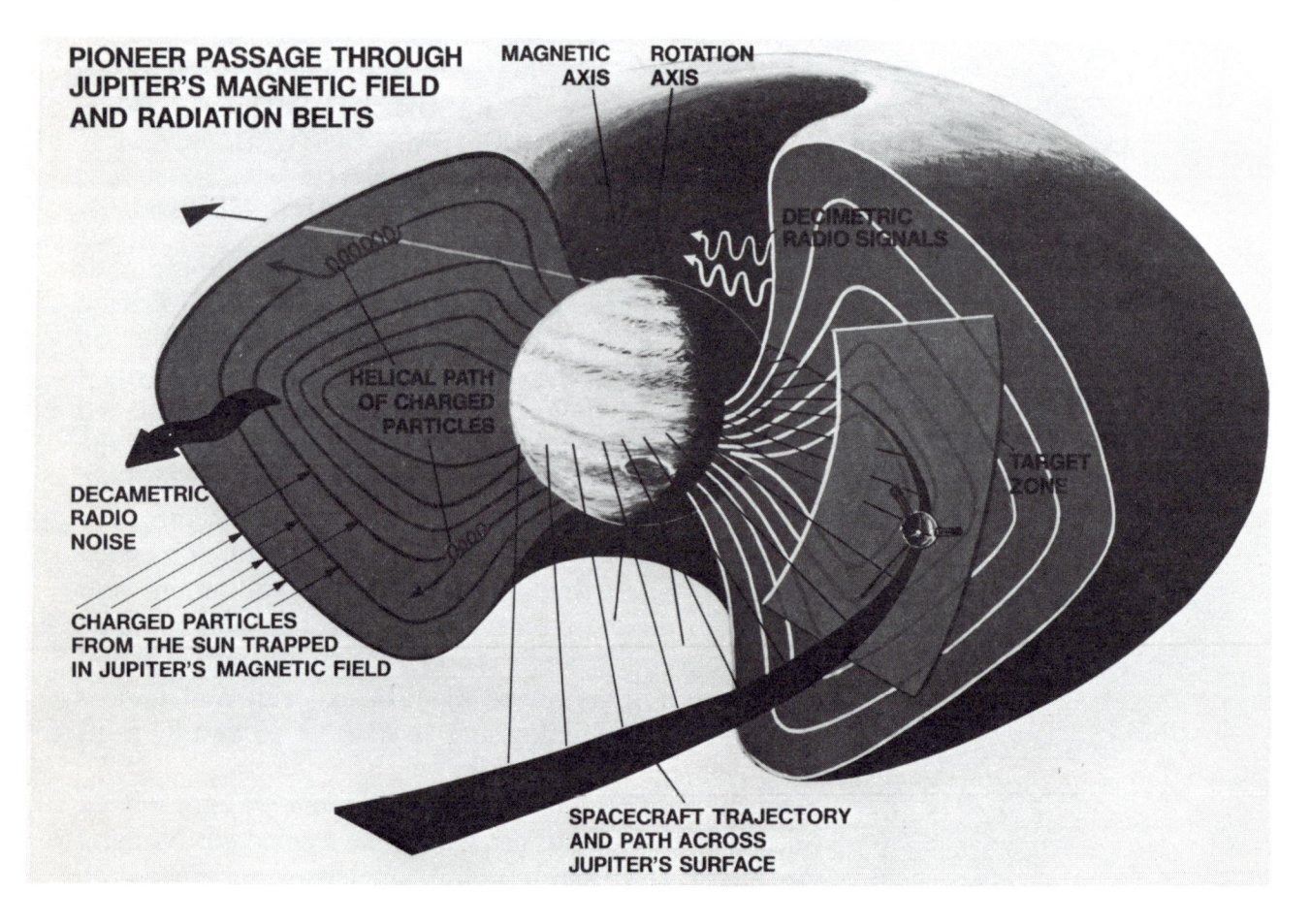

PIONEER PASSAGE THROUGH JUPITER'S MAGNETIC FIELD AND RADIATION BELTS

MAGNETIC AXIS
ROTATION AXIS
DECIMETRIC RADIO SIGNALS
HELICAL PATH OF CHARGED PARTICLES
TARGET ZONE
DECAMETRIC RADIO NOISE
CHARGED PARTICLES FROM THE SUN TRAPPED IN JUPITER'S MAGNETIC FIELD
SPACECRAFT TRAJECTORY AND PATH ACROSS JUPITER'S SURFACE

Structure of Jupiter's magnetosphere, deduced from measurements made by the Pioneer *spacecraft. The magnetic axis is inclined 11° from the Jovian spin axis, and the planet's magnetic poles are quite far removed from its geographic poles.*

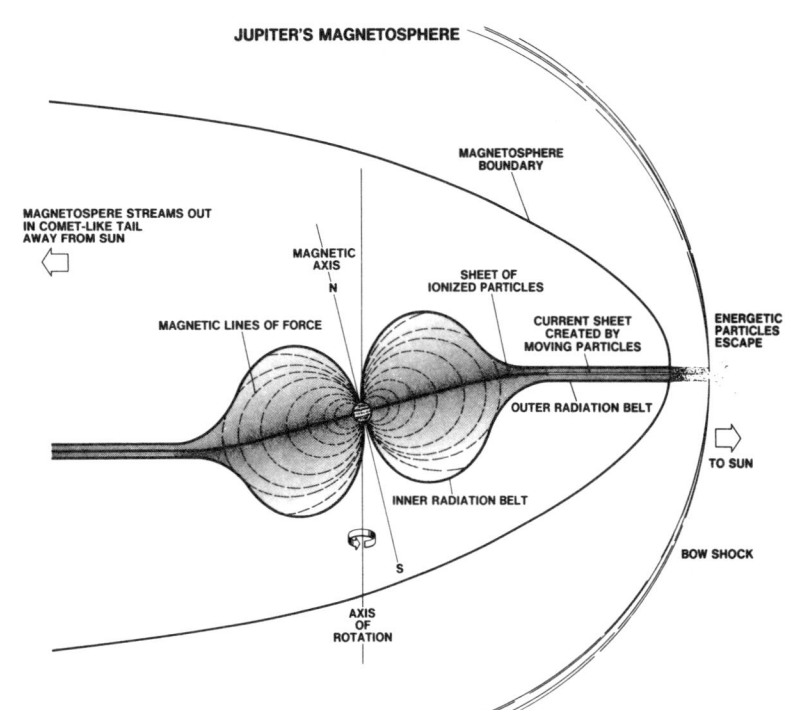

In addition, the communications gear could be used for experimental work; the course of the probe carried it behind Jupiter and, as it happened, behind one of its Galilean satellites as seen from Earth. The radio beam would penetrate various levels of atmosphere (if any) around both bodies, and the effect on transmission would be informative. Further, the motions of the craft itself would be influenced by the gravity fields of all nearby bodies, and give more precise knowledge about masses and even internal mass distribution than centuries of telescope observation could do.

The actual passage of *Pioneer 10* was a tense period, for the radiation was indeed there. Jupiter does have a vast magnetic field, which interacts with the stream of charged particles from the sun known as the solar wind in much the same way that the Earth's field does. *Pioneer 10* ran into the shock front—the boundary region between the volumes of space where wind control is managed by the sun and where Jupiter is in charge—over seven million kilometers from Jupiter, and reported Jovian field conditions for some four days; then it found itself hitting the bow shock again. Apparently a solar event had changed the wind and squeezed the Jovian field region down to a far smaller size. This does not seem to be an unusual event; *Pioneer 10* crossed the front *seventeen times* on its outbound path after passing the planet.

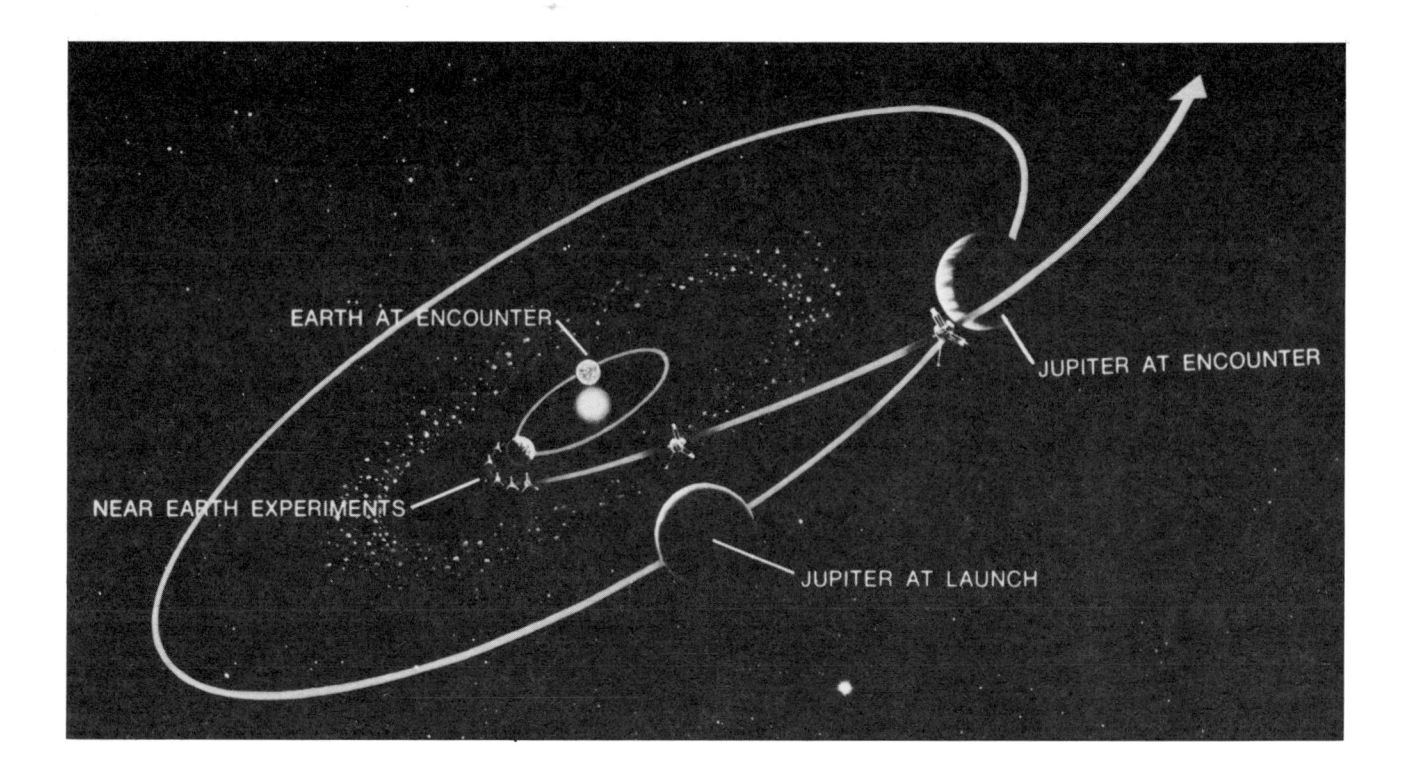

The equipment did survive the hard radiation, though the scientists had their fingers crossed for a while. Vast amounts of information were recorded, and are still being analyzed. Apparently Jupiter's magnetic field is reversed compared to ours—south magnetic pole toward north planetary pole—and tipped 11 degrees to the Jovian axis. It is also located off center in the planet, so different parts of Jupiter's area have very different field strengths and directions. Charged particles are slung around in the whirling field, but centrifugal force tends to move them toward the planet's equatorial plane; hence, the whole sheet of current is rather warped (the word "fedora" has been used to describe it). The giant satellites, especially Io, plow a very noticeable path through the particle belt, strongly affecting the radiation count at various energies.

We have better values for the satellite masses than ever before. For the first time, we can say with near certainty that the densities of the four Galilean satellites vary smoothly with their distance from Jupiter, from little greater than the value of water for the outermost (Callisto) up to three and a half times that value—very like that of our own moon—for Io, the innermost. This is a piece which must somehow be fitted into the jigsaw puzzle of Solar System origin.

The Pioneer 10 *spacecraft was "whipped" past Jupiter by the giant planet's powerful gravitational field, and hurled on a trajectory that will take it out of the solar system and send it toward the distant stars on a journey that will span millions of years.*

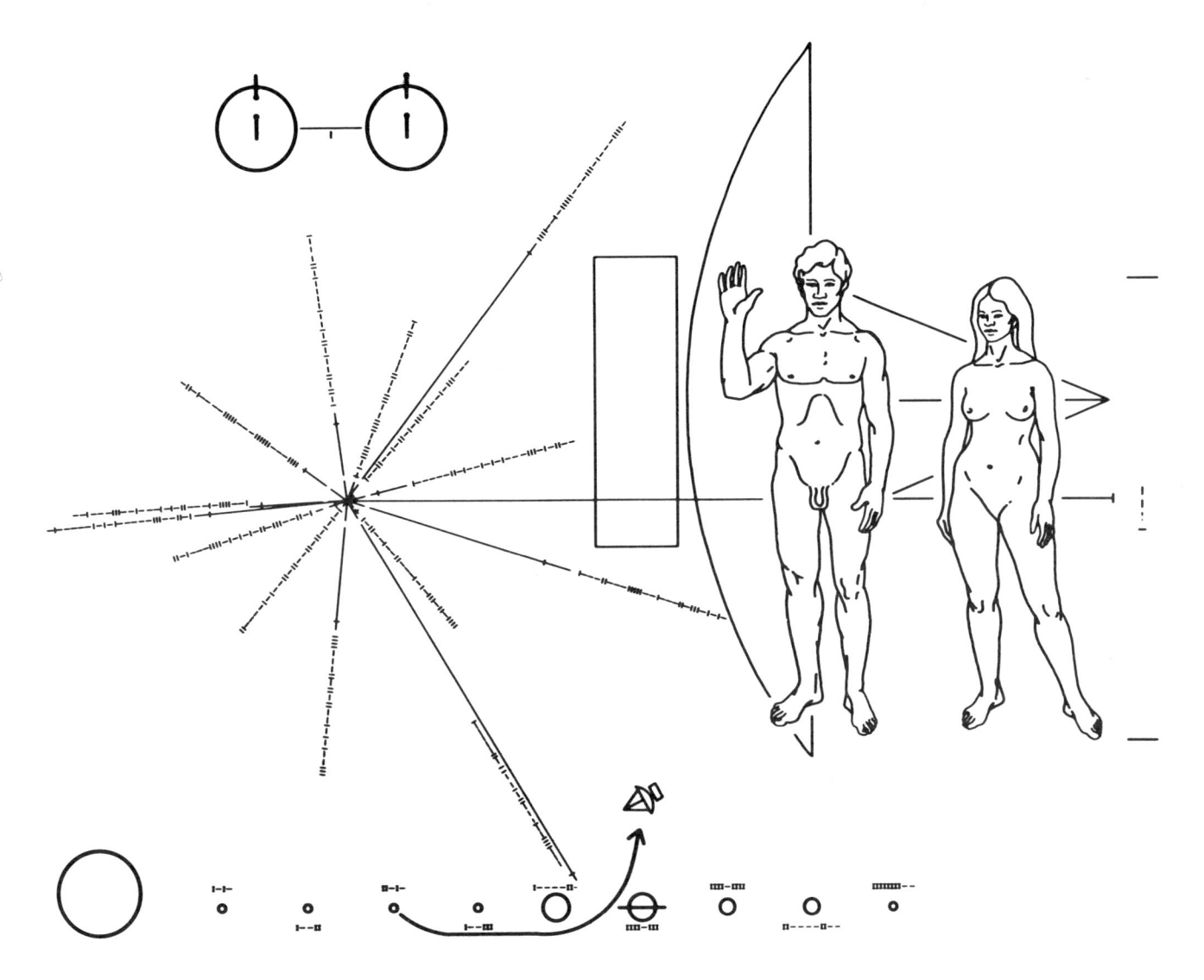

A "message from Earth" to any intelligent alien race that might intercept Pioneer 10 *on its long journey to the stars. This pictorial plaque was attached to the spacecraft, which is the first man-made object to be sent outside our solar system. The plaque is etched into a gold anodized aluminum plate six inches long and nine inches wide (152 by 229 millimeters). The picture is designed to show a scientifically alert extraterrestrial race when the spacecraft was launched, from where, and by what sort of beings.*

Our knowledge of Jupiter's mass has improved slightly—it appears to be three or four thousandths of a percent greater than we had supposed. More immediately interesting, the motion of the spacecraft provided some data on the *internal* distribution of that mass—definitely a new and sharper bit for our mathematical drill. About nine tenths of one percent of the mass is in the outermost 2200 kilometers or so, and the density at that 2200-kilometer depth is about one quarter that of water.

Pioneer 10 actually detected the helium we have long been sure was in Jupiter's atmosphere. It found a faintly glowing doughnut-shaped volume of hydrogen in the space around the giant world's equator. The temperatures above the clouds turned out to be rather higher than had been expected.

Some of the particles which we have been detecting with our Earth satellites and assumed to be from the sun now turn out to be ejected from Jupiter—bursts of

167

electrons powerful enough to travel 80 percent of the way to the sun, along the Solar System magnetic lines of force, *against* the solar wind!

The radio beam grazing Io on its way back to Earth encountered a concentration of ions which makes it possible to say that Io has an atmosphere...

But this must be said very carefully. A couple of decades ago, one of the more sensational and less reliable flying saucer writers made the claim that there was life on all the planets. To support this contention he had a footnote stating that "Dr. Shapley of Harvard Observatory has recently reported the discovery of an atmosphere on the moon." What had really been reported was an occultation event in which the waves from one of the still mysterious radio sources beyond the Solar System had been blocked off temporarily by the moon. As with Io, the presence of charged particles had been shown, indicating a concentration of gas molecules much greater than that in interplanetary space. The moon indeed had an atmosphere—some one ten-billionth as dense as the one you are breathing as you read this. The same general situation exists for Io. If you plan to explore the place, bring your own oxygen along.

All this information is interesting and useful—interesting to the imaginative writer and reader, useful both from the viewpoint of human nature as a curious animal and from the practical need to understand our universe in order to survive in it. However, the most entrancing and fascinating data to come back from the Jupiter probe were the photographs.

The pictures are unreal, in more than one sense. Some extreme purists might say that they included a little deceit, actually. They were taken in pairs, through red and blue filters, and recombined to make single color photographs with a synthetic green signal added. In spite of the artificial additive, there is good reason to feel confident that the resulting pictures show Jupiter very much as it would appear to the human eye at the corresponding distances; and this view is also unreal, in the more commonly used figurative sense.

I am at a loss for adequate words. I am a native New Englander, and while I do tend to take for granted the autumn foliage which is one of the region's tourist attractions, I do go out to look at it. I even yield to the temptation to take color pictures of it, and look at them again later. Sure, it's beautiful.

Jupiter...

The New England colors are there as we look down on the clouds, in detailed patterns as fine as the Pioneer camera could reveal and presumably finer still, just as they are in the Berkshire woods. Intellectually, they may be hydrogen and ammonium polysulfides, just as the leaf colors are carotenes and xanthophylls, but this doesn't detract from the emotional pleasure of looking at them. Reds, browns, yellows, whites and cream colors, blues and greens are carried along the east-west belts and zones by Jupiter's main gas circulation and twisted and knotted into smaller and finer cross patterns by local turbulence. There are white patches which anyone who has seen Apollo photographs of tropical thunderstorms can hardly help interpreting as convection tops; one, a few thousand kilometers north of the equator, shows a streak of white cloud blown westward and a little south for thirty or forty thousand miles—this one shows on many of the pictures; the pattern lasted throughout the several days which elapsed between the start of *Pioneer*'s photography program and the pass a mere 103,000 kilometers above the Jovian clouds.

Jupiter, as photographed from Earth through the 200-inch telescope at Mt. Palomar. The Red Spot is visible, and the moon is Ganymede. (Mt. Wilson and Mt. Palomar Observatories)

I want to see more. If there is anything certain in this life, I would say that every increase in our resolving power is going to make Jupiter an even more beautiful object.

On a less esthetic level, the pictures provide far more detail about the Jovian atmospheric circulation than we have ever had. With over three hundred of these pictures sent back, detailed study of gas behavior has become possible on a scale larger than any laboratory on Earth—larger than Earth itself—could ever provide. I am not myself optimistic about man's ever developing genuine, push-button weather control, in view of the energy scale involved; but if such a thing is ever accomplished, it will be from data of this sort. Whatever happens, these Jovian pictures will add depth to our understanding of the peculiarities of our own atmosphere, negligible as the latter may be compared to the vast gas envelopes of the outer planets.

I remarked a few pages back that it was hard to believe that all the beauty of the Jovian cloudscape could be going to waste, and that there should be intelligent beings there to appreciate it. I was admittedly speaking in a rather tongue-in-cheek fashion; I am not inclined to a teleological philosophy, and I realize that the New England forests were putting on their annual shows long before there were human beings on hand.

Nevertheless, I consider Jupiter as the most likely place in the Solar System for the development of life, not excepting the Earth. I grant that the picture of 40 years ago has had to be modified; the temperature deep in the Jovian bulk is higher than supposed, water as well as ammonia may be liquid, and it is quite likely that there is no true surface in any sense with which we are familiar. Whether the metal-phase hydrogen begins suddenly or is bounded by a few miles or a few hundred miles of syrup where the hydrogen is mixed with hydrogen compounds, I don't know; the quantum mechanics specialists can make calculation on this point, but their figures are extrapolations.

Regardless of this uncertainty, all the chemicals needed for life—the *elements*, that is—are there. We have observed the hydrogen, carbon, and nitrogen, and it is very hard to see how any of the others could be missing from an original sample hundreds of times the mass of the earth. Temperature ranges from below the freezing point of ammonia in the upper atmosphere to well above the melting point of water a few kilometers below the cloud tops. In other words, there is as wide a choice of chemical elements and a broader selection of solvents and solvent mixtures available to any Jovian life than Earth has been able to furnish. The present Jovian environment, in fact, is very much like the one widely believed to have characterized our own planet at the time life started—the obvious differences, I submit, being in favor of Jupiter.

The energy to start the formation of complex life-making compounds is there, in the form of heat and presumably electrical discharges; there is even some ultraviolet in the higher levels, though the hydrogen blocks the really short-wave radiation of this sort. Jupiter, to sum up, is in a better condition than Earth ever was to produce the compounds which we consider necessary for life, and probably to produce others which as far as we know may be just as useful.

If you were an impartial explorer visiting the Solar System in search of life (yes, I realize it's difficult for a living being to be impartial about what is ideal for life, but try) would you look first on a small, almost dehydrated ball of silicates with the barest suggestion of an atmosphere and a temperature range of only a few tens of degrees?

Or would you be attracted to an environment with a really impressive supply of essential materials, plenty of energy, and a respectable range of density, temperature, and pressure?

Be honest, now.

No, Jovian life won't be much like ours. Chemically, the old suggestion of reducing by hydrogen to provide energy for active forms seems reasonable; such reactions can occur. Compounds such as acetylene and cyanogen and acetonitrile and even ethyl alcohol will react with hydrogen to give respectable amounts of energy per pound—less than we get by oxidizing sugar, but not too much less. A really gluttonous Jovian might do an even better job than a human being of growing a spare tire; butadiene, the monomer of rubber, will give about two thirds as much energy per pound reacting with hydrogen as sucrose can get with oxygen...

Mechanically, I would expect ocean-type life, floating or swimming (flying?) free in an environment something like ocean and something like air, with no sharp distinction between the two at the greater depths. Arthur

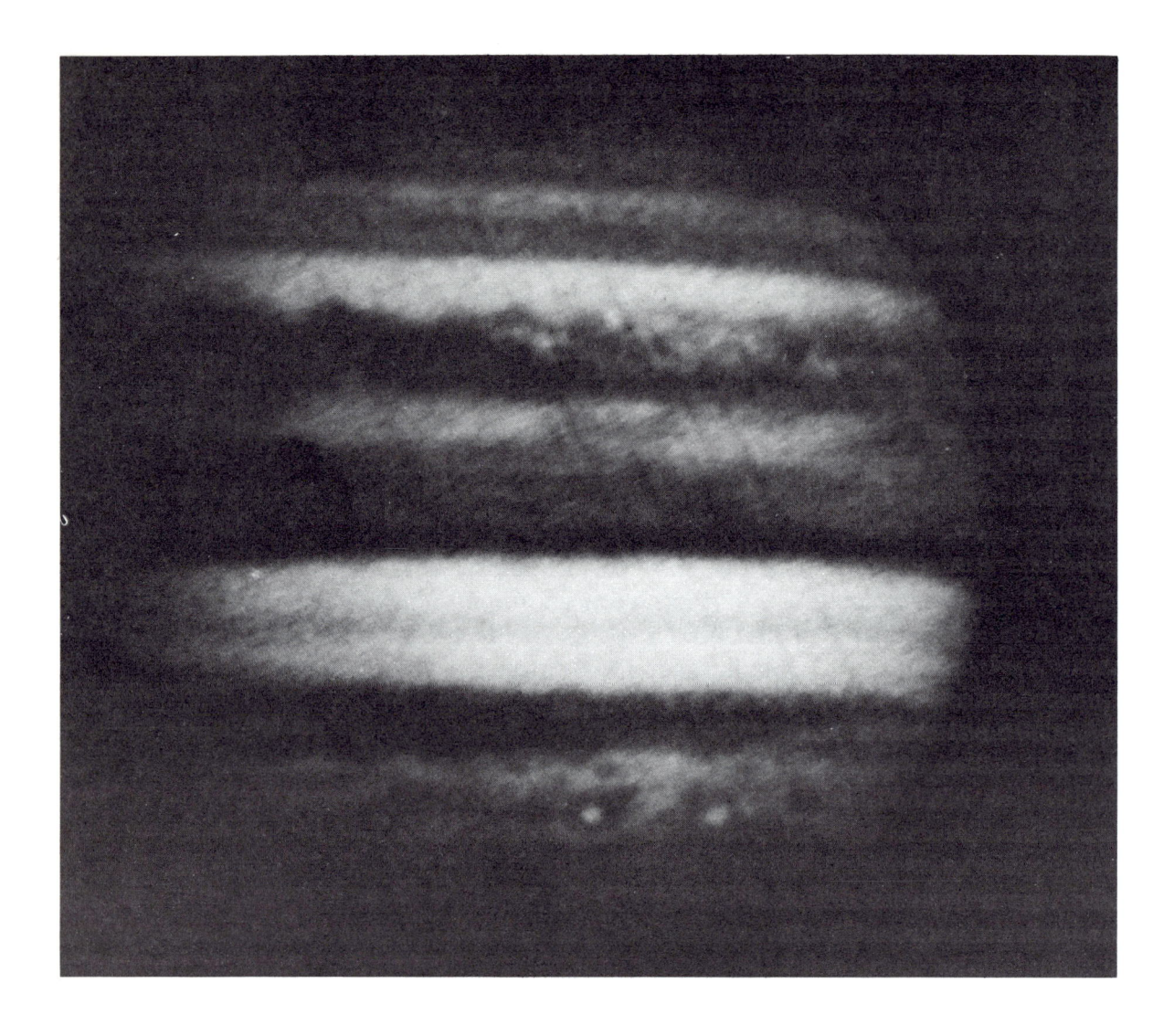

C. Clarke, in his classic "A Meeting with Medusa," probably called the shot pretty closely here.

I am offering this as a serious prediction, not a casual speculation. I hope to live long enough to see it checked. I recognize the present negative attitude toward research in general and space research in particular; many people actually seem to believe that mankind already has enough knowledge to keep the present population alive indefinitely, rather than the few decades we could actually manage. People complain about the "waste" of money spent on space.

I realize that such research is expensive; a single *Apollo* moon flight would keep the population of this country in

Jupiter, as photographed by Pioneer 10 *from a distance of 2,600,000 kilometers. The Earth is never closer than some 740 million kilometers from Jupiter. Note the greatly improved clarity of the view, even though* Pioneer's *cameras were hardly in the same class as the Mt. Palomar facility.*

cigarettes for fully 15 days, and even in liquor for nearly six. I wonder what the Jovians, who may well not know what property lines or cities are, would have for a social system? Maybe the dolphins would understand them—and vice versa—better than we could.

John Campbell, in the article mentioned earlier, spoke of the Jovians looking wistfully and hopelessly out at the other worlds, from which they were forever barred by an environment which could never be built into a space ship. Personally, I doubt that it will ever be the physical environment which keeps an intelligent species from exploring and discovering.

And, since I am still optimistic about human nature in spite of half a century's experience, I feel hopeful about living to see at least a photograph of a Jupiter life form. Intelligence, applied to the real universe, can do some pretty wonderful things.

Editors' Note: Beyond Jupiter, the Solar System is unexplored. True, *Pioneer 11* is travelling out toward ringed Saturn, and will rendezvous with that distant world in 1979. True also that NASA planners hope to send spacecraft toward the even-farther planets of Uranus, Neptune and Pluto.

But to date, all we know of these worlds on the outer fringes of the Solar System is what we have been able to glean from Earth-bound observations. Our understanding of them is as minimal as our understanding of Mercury or Jupiter was before the spacecraft closeups of 1974.

George W. Harper, in this provocative chapter, takes us beyond distant Pluto's orbit, to consider what may lie even further outward—things that no one has seen at all, strange worlds and worldlets that are so far from the Sun that they are too dim to be seen from Earth. This chapter is speculation, entirely; but speculation based on sound knowledge and that one quality that Albert Einstein said was indispensable to scientific progress: imagination.

Styx and Stones; and Maybe Charon Too

George W. Harper

	SATURN
mean radius	58,200 kilometers
mass†	95.2
mean density	.70 grams/cubic centimeter
surface gravity†	1.1
escape velocity	33 kilometers/second
length of day*	10 hours 14 minutes
length of year*	29.46 years
inclination of orbit to ecliptic	2°.5
inclination of equator to orbit	26°.7
mean distance from Sun	9.54 AU = 1.43 x 10^9 kilometers
eccentricity of orbit	0.056

satellites	diameter (km)	distance from planet (millions of kilometers)	orbital period (days)*	date of discovery
Janus	350	0.159	0.75	1966
Mimas	500	0.186	0.94	1789
Enceladus	500	0.238	1.37	1789
Tethys	1000	0.295	1.87	1684
Dione	880	0.377	4.5	1684
Rhea	1500	0.527	4.5	1672
Titan	5800	1.222	15.95	1655
Hyperion	160	1.483	21.28	1848
Iapetus	1700	3.560	79.33	1671
Phoebe	200	12.950	550 *retrograde*	1898

† Earth = 1
* Earth time

URANUS

mean radius	25,400 kilometers
mass†	14.7
mean density	1.21 grams/cubic centimeter
surface gravity†	0.96
escape velocity	22 kilometers/second
length of day*	10 hours 50 minutes *retrograde*
length of year*	84.0 years
inclination of orbit to ecliptic	0°.8
inclination of equator to orbit	97°.9
mean distance from Sun	19.18 AU = 2.87 x 10⁹ kilometers
eccentricity of orbit	0.04

satellites	diameter (km)	distance from planet (millions of kilometers)	orbital period (days)*	date of discovery
Miranda	500	0.130	1.41	1948
Ariel	1600	0.192	2.52	1851
Umbriel	400	0.267	4.14	1851
Titania	~1700	0.438	8.71	1787
Oberon	≥1000	0.586	13.46	1787

NEPTUNE

mean radius	24,500 kilometers
mass†	16.8
mean density	1.7 grams/cubic centimeter
surface gravity†	1.3
escape velocity	25 kilometers/second
length of day*	15 hours 50 minutes
length of year*	164.8 years
inclination of orbit to ecliptic	1°.8
inclination of equator to orbit	28°.8
mean distance from Sun	30.07 AU = 4.497 x 10⁹ kilometers
eccentricity of orbit	0.009

satellites	radius (km)	distance from planet (millions of kilometers)	orbital period (days)*	date of discovery
Triton	~2900	0.355	5.88 *retrograde*	1846
Nereid	150-250	5.562	360	1950

† Earth = 1
* Earth time

PLUTO

mean radius	5800? kilometers
mass†	0.11?
mean density	0.49? grams/cubic centimeter
surface gravity†	?
escape velocity	5.3 kilometers/second
length of day*	6.25 days
length of year*	248.4 years
inclination of orbit to ecliptic	17°.2
inclination of equator to orbit	?
mean distance from Sun	39.44 AU = 5.90 x 10^9 kilometers
eccentricity of orbit	0.249

satellites	diameter (km)	distance from planet (millions of kilometers)	orbital period (days)*	date of discovery
unknown	NA	NA	NA	NA

† Earth = 1
* Earth time

In 1766 the German mathematician Johann Titius wrote a brief footnote to a book on natural philosophy he was translating from the French. The book itself is long forgotten save for a few scholars, but the footnote has led a lively career. As translated by Stanley L. Jaki of Seton Hall University, it reads: "Divide the distance from the sun to Saturn into 100 parts; then Mercury is separated into 4 such parts from the sun; Venus by 4 + 3 = 7 such parts; the Earth by 4 + 6 = 10; Mars by 4 + 12 = 16. But notice that from Mars to Jupiter there comes a deviation from this exact progression. After Mars there follows a distance of 4 + 24 = 28 parts, but so far no planet or satellite has been found there... Let us assume that this space without a doubt belongs to the still undiscovered satellites of Mars... Next to this for us still unexplored space there rises Jupiter's sphere of influence at 4 + 48 = 52 parts; and that of Saturn at 4 + 96 = 100."

Nor was this the first prediction of a planet between Mars and Jupiter. Nearly two centuries earlier, around 1595, Johannes Kepler penned the unambiguous sentence: *Inter Jovem et Martem planetum interposui*, or "Between Jupiter and Mars I interpose a planet."

Either way, the mathematical relationship expressed by Titius and the prediction by Kepler remained curiosities until the summer of 1781, when William Herschel proclaimed a new planet in the firmament, a planet which he named "Georgium Sidus" in honor of King George III, of Revolutionary War fame. With this discovery it was quickly realized the new planet fitted neatly into the next interval of the Titius Rule, at 4+ 192 = 196.

Herschel's discovery refocused attention on the Titius Rule and incidentally, on a rather nasty but unfortunately somewhat typical situation which had arisen during the previous 15 years. The preeminent German astronomer of the time, one Johann Bode, had simply appropriated the Titius Rule and claimed it for his own despite the fact he had earlier explicitly acknowledged Titius' priority. As he had the wholehearted cooperation of the German astronomical fraternity, the expropriation stuck and it is today generally known as the "Bode" Rule.

Ordinarily we wouldn't mention Bode here save for an unexpected irony. Bode was the astronomer who renamed Georgium Sidus, calling it Uranus. So, oddly enough, he got credit for the rule he stole and never received proper recognition for the planet he named! Maybe things come out even after all!

Still, the rule remained mostly a curiosity until New Year's eve, 1801, when the astronomer-monk Giuseppe Piazzi discovered what he first believed to be a peculiar comet with an unusually circular orbit in a position roughly between Mars and Jupiter. But then Karl Friedrich Gauss proved it to be a small planet, orbiting at a distance of 27.7. The planetoid was later named Ceres and has since been proven to be the largest of the asteroids.

Once the asteroids were fitted into the 28 slot, the Titius Rule stood triumphantly confirmed. As of 1801 the Solar System had a neat, complete look about it. Everything was in its place and all was right in the heavens. The *reason* for the rule might be obscure, but the *reality* was unquestioned.

Shortly after the discovery of Uranus, astronomers

began to realize that the existence of the planet could have been predicted far in advance of its discovery. The period of Uranus is 84 years. The period of Saturn is only 29½ years. This means that roughly every 40 years Saturn and Uranus come into conjunction. When Saturn starts to overtake Uranus it is accelerated by the gravitational attraction of the outer planet. When it passes out of conjunction and begins receding, the attraction of Uranus pulls on Saturn and slows it down.

The actual effect of this is the precise *opposite* of what we should expect. The acceleration of Saturn toward Uranus translates itself into a higher orbit and a consequent reduction of speed in its motion about the sun. As it passes Uranus, the gravitational drag is converted into a lower orbit and an increase in speed. The fits and starts of Saturn had been observed for years without anyone ever suspecting the reason, but once the phenomenon was recognized, mathematicians commenced analyzing the motions of Uranus and Saturn to look for evidence of additional residuals which might indicate the presence of other planets.

They found one.

John C. Adams in England and U. L. J. Leverrier in France arrived independently at the same conclusions. Adams was a bit ahead of his rival, but he made the mistake of turning his calculations over to the Astronomer Royal of England. And that worthy had better things to do with his time than worry about the calculations of some amateur. Leverrier had better luck, and on September 23, 1846, Neptune was discovered.

Then a ripple of dismay began spreading through the ranks of astronomers. Rather than falling at 388 as the Titius Rule suggested, Neptune orbited at a scant 300, or over 1,300 million kilometers from where it belonged!

When Pluto was discovered, some 84 years later, the difficulty was compounded. Rather than orbiting at 772, as the Titius Rule predicts, it loops out in a highly eccentric orbit ranging from 290 to 420, and averaging 394. (See Table I.)

In other words, after Uranus the whole system goes to pot! One result of this has been an effort by some astronomers to call the whole rule a fluke. "Pure coincidence," they scoff... but even in their scoffing we sense a certain uneasiness; as if there is a lurking fear there is unfathomed significance to the old rule after all.

TABLE I. (Titius Rule)

planet	Titius interval	true value
Mercury	0.4	0.39
Venus	0.7	0.72
Earth	1.0	1.00
Mars	1.6	1.52
Asteroids (Ceres)	2.8	2.77
Jupiter	5.2	5.20
Saturn	10.0	9.54
Uranus	19.6	19.18
Neptune	38.8	30.06
Pluto	77.2	39.44

The fear seems justified when we turn to look at the satellite systems circling some of the outer planets. Take Uranus, for instance. Here we find five beautiful satellites, all in perfect equatorial orbit about the planet, and all with very nearly zero eccentricities. If we apply the Titius Rule here, we find an excellent approximation save for a moderately considerable discrepancy with the innermost satellite, Miranda, and a massive discordancy with the outermost, Oberon. (See Table II).

When we consider the five inner, regular satellites of Jupiter, also listed in Table II, we again arrive at an interesting approximation of the rule. Amalthea and Io are definitely too close to their primary, but Ganymede and Callisto are squarely on the mark. The more distant satellites, being highly eccentric and inclined in orbit, are considered to be later acquisitions and thus not subject to the rule.

TABLE II. (Satellite systems)			
planet & satellite	Titius interval	true value	actual distance (in Km)
Uranus			
5. Miranda	0.4	0.487	130,100
1. Ariel	0.7	0.720	191,800
2. Umbriel	1.0	1.000	267,300
3. Titania	1.6	1.640	438,700
4. Oberon	2.8	2.195	586,600
Jupiter			
5. Amalthea	0.4	0.27	181,000
1. Io	0.7	0.63	421,800
2. Europa	1.0	1.00	671,400
3. Ganymede	1.6	1.60	1,071,000
4. Callisto	2.8	2.81	1,884,000

A substantial improvement in the accuracy of the Titius Rule is achieved if we postulate that whenever a given condition is fulfilled at the outer edges of the system, the planets or satellites out there will tend to condense at half intervals. The precise nature of this condition is unimportant at the moment, but as a guess we may hazard it is somehow related to the density of matter per unit volume of space. But even if the reason is obscure, the fact of the improvement is real. Neptune and Pluto fall neatly into place in the Solar System, and Oberon fits just as neatly into the pattern of Uranus' satellites. (See Table III.)

TABLE III. (Modified Titius Rule)

planet		Titius formula	true value
Mercury		4 + 0 = 0.4	0.39
Venus		4 + 3 = 0.7	0.72
Earth		4 + 6 = 1.0	1.00
Mars		4 + 12 = 1.6	1.52
Asteroids (Ceres)		4 + 24 = 2.8	2.77
Jupiter		4 + 48 = 5.2	5.20
Saturn		4 + 96 = 10.0	9.54
Uranus		4 + 192 = 19.6	19.18
Neptune	(Standard)	4 + 384 = 38.8	
	(Modified)	4 + 288 = 29.2	30.06
Pluto	(Standard)	4 + 768 = 77.2	
	(Modified)	4 + 384 = 38.8	39.44
X	(Modified)	4 + 576 = 58.0	?
X + 1	(Modified)	4 + 768 = 77.2	?
Uranus satellites		Titius formula	true value
5. Miranda		4 + 0 = 0.4	0.487
1. Ariel		4 + 3 = 0.7	0.720
2. Umbriel		4 + 6 = 1.0	1.000
3. Titania		4 + 12 = 1.6	1.640
4. Oberon	(Standard)	4 + 24 = 2.8	
	(Modified)	4 + 18 = 2.2	2.195

It may be objected that the failure of the Saturn satellites to conform invalidates the hypothesis, but we may counter by observing that the Saturn family is exceptional in more ways than just this one. For instance, how do we account for the fact that Mimas and Enceladus have abnormally low densities, being only 0.5 and 0.7 that of water respectively? Tethys, the fourth satellite out, has a density of 1.2, and we can show that the small size of the Kirkwood gaps in Saturn's rings precludes a density greater than 0.4 for Janus, the innermost satellite of Saturn. In fact, not until the fifth satellite, Dione, do we begin to develop 'normal' satellite densities. (See Table IV.)

Conventionally, astronomers draw a distinction between "terrestrial" and "jovian" planets and satellites, calling Saturn's inner family "jovian." It seems likely this is an artificial distinction, especially since Jupiter has no "jovian" satellites and we find no evidence of a pattern in the placement of these satellites around other planets. It seems more probable the same factors which contributed to the formation of the ring system also messed up the

TABLE IV. (Saturn's family)

planet or satellite	distance	radius (km)	density	inclination	eccentricity
Saturn		60,400	0.687		
Janus	159,200	240	0.4	0.0 ?	0.004 ?
Mimas	185,700	250	0.5	1.5	0.0196
Enceladus	238,200	320	0.7	0.0	0.0045
Tethys	294,800	510	1.2	1.1	0.0000
Dione	377,700	450	2.8	0.0	0.0021
Rhea	527,500	650	2.3	0.3	0.0009
Titan	1,203,000	2401	2.42	0.3	0.0289
Hyperion	1,484,000	200	1.58	0.6	0.110
Iapetus	3,563,000	600	2.21	14.7	0.290
Phoebe	12,950,000	150	?	210.0	0.166

Saturn's Rings

Outer Radius from Saturn's center = 138,000 km.
Outer A Ring: (Bright) = 120,000 km.
Cassini Division (Dark) = 116,000 km.
Main B Ring: (Very Bright) = 90,000 km.
Inner (Kirkwood) Gap: (Dark) = 89,000 km.
Crepe Ring: (Very Faint) = 71,000 km.
Surface of Saturn: = 60,400 km.
Thickness of Rings = 15 km.
Mass of Rings = 0.00004 the mass of Saturn, or 1/2,375,000 the mass of Earth

Titius Rule and created a whole set of underdense satellites with anomalous orbits.

Admittedly, the argument is not overwhelmingly convincing, but with so many peculiarities in and around Saturn, we need not be surprised when the Titius Rule also goes by the wayside. It is simply one more oddity in the system.

So in summary, it looks as if the Titius Rule contains elements of reality and represents something more than simple coincidence. Granting this much, if a tenth planet should exist in our Solar System we would expect to find it wandering in orbit at around 58 astronomical units. An eleventh planet would probably fall somewhere around 77.2 a.u. Further, as the formula for naming planets is already fairly well established, we can go ahead and name a tenth planet "Styx" and an eleventh "Charon" without doing violence to tradition.

This is fine as far as it goes, but there is a fly in the ointment... Pluto. Considering the true value for Neptune and the half intervals of the modified Titius Rule, Pluto is exactly where it belongs. But this is almost the *only* thing right about the planet. Everything else is wrong. Its orbit is too eccentric, its mass is too small, its composition and density evidently wrong, and the rotational period faulty. In short, astronomers would probably much prefer that Pluto were not around. But unfortunately, it is there, and we have no convenient way of ignoring the planet. So we must try to explain it.

The matter begins in 1915, when Percival Lowell published an expertly developed mathematical analysis of observed deviations in the orbit of Uranus. From these he deduced the existence of a planet beyond Neptune and arrived at a probable location.

But this was not the first effort to seek out a transneptunian planet. As early as 1834, the Danish astronomer Peter Andreas Hansen indicated a belief that a single planet would not account for the residuals in the orbit of Uranus. In 1880, David P. Todd made a systematic search using the 26-inch refractor at the U.S. Naval Observatory. There were others too, but these were probably first in their respective areas. Hansen first suggested the planet, Todd first sought for it, and Lowell first arrived at a mathematical prediction.

When Pluto was finally discovered, in March of 1930, it turned out to be within six degrees of Lowell's predicted position. This is phenomenally good mathematics, and the likelihood of coincidence is negligible. But even as the discovery was being announced, astronomers at the Lowell Observatory were hedging their comments. If Lowell's mathematics were correct, Pluto had to have a mass 6.6 times that of Earth; that is, assuming the distance at which it was actually found. This led to problems, for the planet appeared to be about the size of Mars, or roughly .25 the volume of Earth. This would imply a density of 147.0 for Pluto as contrasted to Earth's density of 5.52, which would make Pluto consist mainly of collapsed matter!

Unfortunately, this creates its own problems. It happens Pluto's orbit is the most eccentric of any planet in the system. At its point of nearest approach to the sun, on May 5, 1989, it will actually be located *within* Neptune's orbit! As Pluto's orbit is highly inclined, with respect to Neptune's, there is no danger of collision with Neptune, but it is a lead-pipe cinch any planet with 6.6 Earth masses coming that close to Neptune would perturb it mightily over the ages, both in terms of orbital ellipticity and inclination. And what do we find? We find Neptune to be second *least* perturbed of any planet in the Solar System, with an eccentricity of 0.0087. Venus is slightly better, with an e of 0.0068, while Earth is just behind, with an e 0.0167, roughly twice as great. Uranus is a distant fourth, with an e six times larger than Neptune's. This datum alone causes all other arguments to pale to insignificance. There is simply no way Pluto can wind up with a highly considerable mass. The assumption of .10 Earth mass for Pluto seems about right, and it is difficult to concede any more.

If we accept all this, it means Lowell's mathematics were accidental. The mass of Pluto turns out to be so inconsiderable there is no way it could give results of the magnitude postulated. Taken at face value, the whole discovery becomes a fluke...or so goes the argument today.

Taken by itself, the matter could easily be dismissed. After all, Pluto is still in the right place so far as the Titius Rule is concerned. If more planets are to be found we should expect to find them at 58.0 and 77.2 a.u., so it really makes little difference if Pluto turns out to be smaller than we anticipated. This would appear to be a clear and concise conclusion.

But there is a problem. In Pluto we have a tiny planet with an orbit intersecting that of a major planet. The question inevitably arises, how stable can such orbit be? Is there perhaps a point in time where the two would have to bump?

Computer simulations fail to reveal such a point, but as they can only be projected a few millions of years into the past and future, this is inconclusive. Much can happen in four billion years which wouldn't even be hinted at in the course of a few million. Thus, there is a distinct possibility of collision, either in the past or the future.

A collision in the future is simply an interesting possibility. Almost certainly the last, enfeebled descendants of humanity will have long since perished ere this time comes. And there is no conceivable connection with our problem at the moment. Whether or not Pluto collides with Neptune is irrelevant so far as the Titius Rule is concerned.

But there is an unexpected relevancy when we look to the distant past. There is a distinct possibility Pluto is not properly a planet at all, that it is instead an escaped satellite of Neptune! And if so, then we don't really have a planet to put into the 38.8 a.u. slot.

Impressive evidence supports the thesis. First is the fact that Pluto's probable radius of 2,650 kilometers is on the same order as Titan, Ganymede or Callisto. It is only slightly larger than the 2,000-kilometer radius of Neptune's major satellite, Triton. The size is therefore about right for a satellite.

Then comes another peculiarity, the rotation period of the planet. Being so far from the sun, tidal effects would be negligible and any planet would retain its aboriginal spin unchanged over eons of time. Thus Jupiter spins once ever 9 hours, 50 minutes. Saturn takes 10 hours, 14 minutes, Uranus 10 hours, 49 minutes, and Neptune 15 hours, 48 minutes. Then comes Pluto with an absurd period of 6.39 days!

Clearly, something had to slow it down, and that can only have been some sort of tidal effect operating somewhere. The only visible way of providing a drag of this magnitude is to assume Pluto was once a satellite of Neptune in a 6.39-day orbit. Then the period would be synchronous with the rotation and our problem would be solved.

Perhaps the most impressive bit of evidence in support of this thesis is Neptune's major moon, Triton. It looks quite normal, as moons go. The radius of 2,000 kilometers is a bit large, but not exceptionally so. The eccentricity of the orbit is zero to four decimal places, which makes it as nearly perfect as possible. The period about Neptune is a nice 5.88 days and its orbital distance from the planet is 353,600 kilometers.

But now comes the clinker... Triton travels *backward* in its orbit around Neptune!

Admittedly, there are a few other satellites which go the wrong way around their primaries. Jupiter has four retrograde satellites, having radii of 10, 9, 9 and 8 kilometers respectively. Their eccentricities are all greater than 0.13, or at least 13,000 times greater than Triton while their diameters are on the order of 1,000 times less.

Saturn adds one more to the collection. Little Phoebe is a moonlet with a diameter of 120 kilometers and an *e* of 0.166. It is also nearest of the other retrogrades to its primary, being a mere 13 million kilometers from Saturn, or some 30 times further out than Triton.

In short, the other retrogrades are small, highly eccentric in orbit and very distant from their primaries. Current belief is they were all captured at some time in the past. But the possibility of Triton having been captured is so slight as to be virtually nonexistent. We are therefore left with the inadmissible conclusion it must have formed *in situ* around Neptune only traveling backward in orbit. Clearly, there is a need for an alternative choice.

R. A. Lyttleton of Cambridge University put it all together. He began with the assumption that Pluto was originally a satellite of Neptune in 6.39-day orbit some 500,000 kilometers distant. Triton was also a regular satellite of Neptune in normal orbit at perhaps 600,000 kilometers. Gravitational interaction caused the two satellites to converge until eventually Triton and Pluto whipped about one another in near collision, with Triton winding up in a lower, circular, retrograde orbit about Neptune while Pluto was cast off as a runaway satellite.

The orbit of the ex-satellite would naturally reflect its point of origin so we would expect it to have a perihelion close to Neptune's orbit. Further, as Triton and Pluto would have a common birth in their present configuration, we would also expect them to have similar inclinations in their respective orbits. And sure enough, Pluto has an exceptionally high 17°.13 inclination, far higher than any other planet in the system. Triton matches this with an inclination of 20°.10±2°.3. Subtract Neptune's own inclination of 1°.77 and we arrive at a relative value of 18°.33±2°.3 for Triton; phenomenally close to Pluto. Lastly, nothing in all this would change the rotational angular momentum of Pluto itself, so the new planet would continue to possess a 6.39-day rotation period as a memento of its dependence on Neptune.

All in all, this is a convincing argument. Everything winds up being explained in terms of simple, easily understandable mechanics. If correct, Pluto does not belong as the ninth planet. It simply chanced to get there by accident. And if this is the case, then for there to be planets at 58.0 and 77.2 would imply the existence of some planet *other than Pluto at approximately the mean orbital distance of Pluto!* It would have to be this as yet undiscovered planet which fills the Titius Rule slot at 38.8.

This is not an impossible requirement. The sidereal

Comet Mrkos, photographed in 1957 with the 49-inch Schmidt telescope at Mt. Palomar. (Mt. Wilson and Mt. Palomar Observatories)

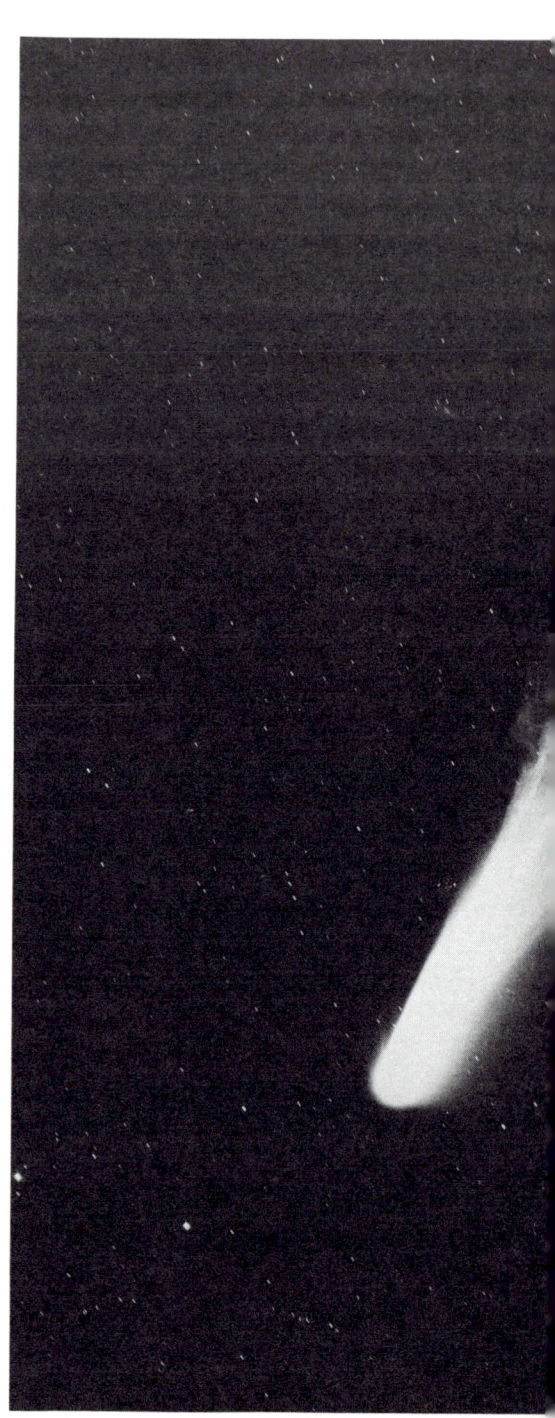

period of a planet in orbit at 38.8 is in the neighborhood of 250 years. In the 42 years since discovery we have observed Pluto over only 1/6 of a single orbit. There is therefore a distinct possibility another planet could exist in the same approximate orbit as Pluto without our having discovered it. There are, after all, some thousands of minor planets in the asteroid belt, so another planet at 38.8 is by no means out of the question. The chances are it would not be more than twice Pluto's diameter or it would have shown up in the extensive planet searches sponsored by the Lowell Observatory, but this would still make it nearly terrestrial in size and mass, so it would be no mean object.

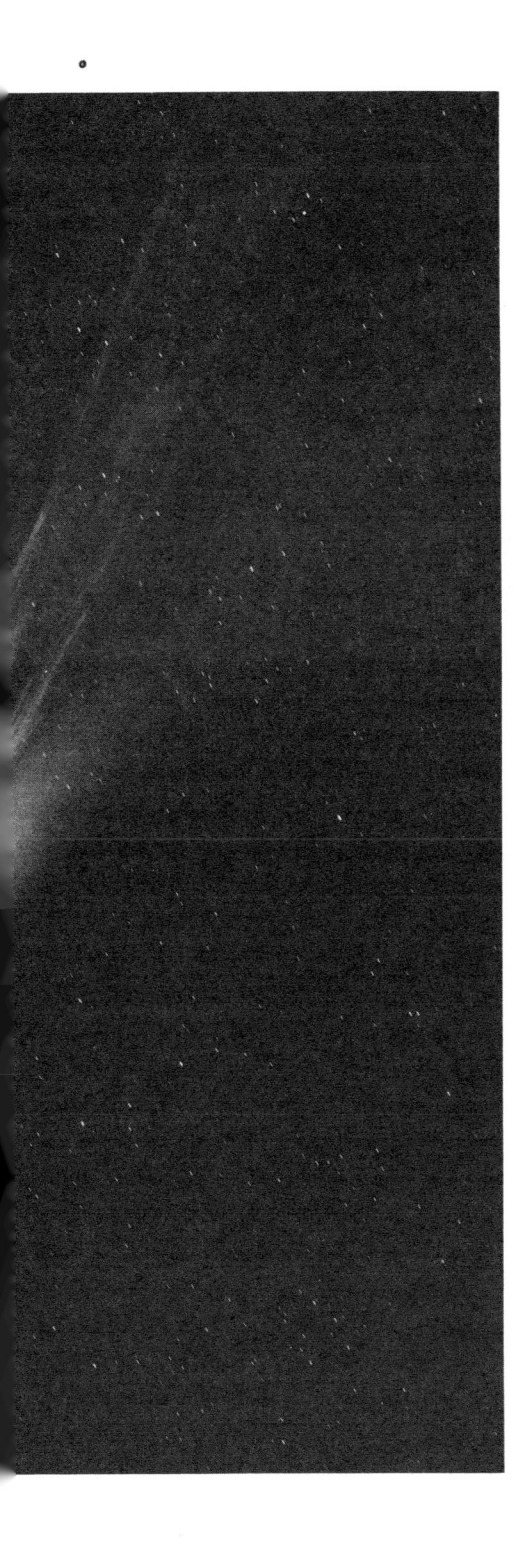

When we start talking about the likelihood of such a planet, that becomes a different matter. The extensive searches by Clyde Tombaugh (who discovered Pluto) make it appear unlikely, but he by no means blinked all segments of the heavens, so there is a reasonable possibility such a planet might exist. If it is as small or smaller than Pluto, and at a distance of 38.8 a.u., there is a good chance it would have been missed even on a direct search. (Pluto was about 34 a.u. from the sun when it was discovered; a planet of the same size at 39 a.u. would be 740 million kilometers more distant and only about half as bright.) But this isn't the point. We have no right to postulate extra planets just for the fun of it. There should be some real reason or we are simply playing games and it becomes an exercise in airy speculation. So we must ask if there is some empiric reason to postulate one or more extra planets at and beyond Pluto.

This is a difficult question to answer. For example, it is entirely possible to explain away the disturbances in Uranus' orbit in terms of inaccurate early observations mated to highly accurate later ones. Thus the residuals Lowell used would all be imaginary and there would be no significance to the mathematical results he achieved. We could even argue there was a positive emotional push for astronomers of the last century to interpret any vagrant residual as evidence of more distant, undiscovered planets. The successes of Herschel, Adams and Leverrier testified to the honors awaiting the discoverer of a new planet, and ambitious astronomers were eagerly seeking ways of joining the select group.

This is the argument being advanced today by those who feel Lowell's calculations were merely a lucky chance. We admit the strength of the argument. But we must also note that the modern pressure is in precisely the opposite direction. The young mathematician of today scurries around in the mathematics of Lowell and others, picking up a residual here, another residual there, and tacks them all together in the presence of "uncertainties," and finally pronounces that he can explain Lowell's "error."

Of course, all he has done is assume that all errors accumulated over the years were "positive" with no "negative" errors to balance. This is highly unlikely. The thought of competent observers over a stretch of two centuries all making the same sort of error in total ignorance of each other boggles the imagination. It just isn't likely. Lowell's computations retain a definite attraction. No matter how cavalierly dismissed, there remains a powerful suspicion he said something worth listening to. And if so, at least one more planet must exist beyond Neptune.

Joseph L. Brady and Edna Carpenter, of the University of California's Lawrence Radiation Laboratory, postulate a planet of 300 Earth masses orbiting at 59.9 a.u. and inclined 120° from the ecliptic. They derive these values from observed discrepancies in the return of Halley's Comet as reported from A.D. 295 to the present.

Unfortunately, a direct scan and blink comparison of the predicted location fails to disclose a planet. Further, rediscussion of the apparitions of Halley's Comet tends to throw doubts on the dates adopted by Brady, so here again it looks like a standoff.

But there is still another line of argument; one used by Brady but still not made explicit by him. This argument derives from the theory of comet "families." Going back a bit, the best evidence today suggests that the entire Solar System is englobed by a cometary "halo," consisting of some 50 million comets in slow orbit about the sun at distances ranging from 30,000 to 50,000 a.u. This works out to perhaps one cometary mass for each volume of space equal to a sphere with a radius the size of Earth's orbit about the sun.

Occasionally one of these bodies interacts with another and both are perturbed out of their circular orbit. If the perturbation is less than escape velocity for the system, both bodies are fated ultimately to plunge inward toward the sun in a long, elliptical orbit. Generally these orbits are so eccentric the comet will have a period running into the millions of years.

But if the circumstances are just right, at some point on its inward plunge or outward return to the depths of space the comet will be perturbed by one of the planets, such as Jupiter. When this happens, the period of the comet is shortened and it becomes a reflection of the period of the perturbing planet. Thus we have the jovian family of comets, having periods of 10 years or less, a Saturn family with periods ranging from 10 to 20 years, a Uranus family of 20 to 40 years, and a Neptune family of 40 to 100 years. According to Table V, 39 comets belong to the jovian family, six to the Saturn family, three to Uranus and five to Neptune. *Then there are two others with periods which appear consistent with a planet at 58.0 a.u.!*

Actually, we can probably add three more comets to the 58.0 family. These are Swift-Tuttle, found in 1862, with a period of 119.9 years; Barnard (2), found in 1889, with a period of 145.4 years; and Mellish, discovered in 1917, with a period of 145.4 years. The comets on Table V have all been observed through more than one apparition and so have fairly reliable orbits established, but these three have been observed only once apiece and have somewhat doubtful orbits. It is unlikely that any of these three has had its orbit so badly misjudged as to be completely out of the area, so we can probably feel fairly safe in attributing five comets to the 58.0 a.u. family.

But like almost everything in astronomy, we can argue with the conclusions. Objectors to the idea of comet families point to the high inclinations of such objects as Halley's Comet and argue that Neptune could not possibly be of significance in modifying the orbit. They maintain that the real culprit for virtually all periodic comets is Jupiter. They further maintain that blocking off decades of time and claiming some sort of mysterious connection with the planets is mere numerology.

On the balance, this is one place where the argument of the objectors is clearly the stronger of the two. There is no doubt that the theory of comet families is correct as stated, but the application should almost certainly be restricted to comets on approximately the same plane as the planets influencing them. Halley's Comet, for instance, passes within 4.6 a.u. of Jupiter, but it never comes within 25 a.u. of Neptune. Clearly, the influence of Jupiter will be vastly the greater of the two. For that

TABLE V. (Cometary orbits)

COMET	Date Last Seen	Period in Years	Perihelion Distance (AU)	Eccentricity	Aphelion Distance
Encke	1974.32	3.30	0.338	0.847	4.09
Grigg-Skjellerup	1972.17	5.12	1.001	0.663	4.94
Tempel 2	1972.87	5.26	1.364	0.549	4.68
Honda-Mrkos-Pajdusakova	1974.99	5.28	0.579	0.809	5.49
Neujmin 2	1927.04	5.43	1.338	0.567	4.84
Brorsen	1879.24	5.46	0.590	0.810	5.61
Tempel 1	1972.54	5.50	1.497	0.520	4.73
Tuttle-Giacobini-Kresak	1973.41	5.56	1.152	0.633	5.13
Tempel-Swift	1908.76	5.68	1.153	0.638	5.22
Wirtanen	1974.51	5.87	1.256	0.614	5.26
D Arrest	1970.38	6.23	1.167	0.656	5.61
Du Toit-Neujmin-Delporte	1970.77	6.31	1.677	0.509	5.15
De Vico-Swift	1965.64	6.31	1.624	0.524	5.21
Pons-Winnecke	1970.55	6.34	1.247	0.636	5.61
Forbes	1974.38	6.40	1.533	0.555	5.36
Kopff	1970.75	6.41	1.567	0.546	5.34
Schwassmann-Wachmann 2	1974.70	6.51	2.142	0.386	4.83
Giacobini-Zinner	1972.59	6.52	0.994	0.715	5.98
Wolf-Harrington	1971.67	6.55	1.622	0.537	5.38
Biela	1852.73	6.62	0.861	0.756	6.19
Tsuchinshan 1	1971.71	6.64	1.493	0.577	5.57
Perrine-Mrkos	1968.84	6.72	1.272	0.643	5.85
Reinmuth 2	1974.35	6.74	1.941	0.456	5.19
Borrelly	1974.36	6.76	1.316	0.632	5.84
Johnson	1970.24	6.77	2.200	0.385	4.96
Tsuchinshan 2	1971.91	6.80	1.775	0.505	5.40
Harrington	1960.49	6.80	1.582	0.559	5.60
Gunn	1976.11	6.80	2.445	0.319	4.74
Arend-Rigaux	1971.26	6.84	1.444	0.599	5.76
Brooks 2	1974.01	6.88	1.840	0.491	5.39
Finlay	1974.50	6.95	1.096	0.699	6.19
Holmes	1972.08	7.05	2.155	0.414	5.20
Daniel	1964.30	7.09	1.662	0.550	5.72
Harrington-Abell	1969.36	7.19	1.773	0.524	5.68
Shajn-Schaldach	1971.75	7.27	2.227	0.406	5.28
Faye	1969.77	7.41	1.616	0.575	5.98
Ashbrook-Jackson	1971.20	7.43	2.285	0.400	5.33
Whipple	1970.77	7.47	2.480	0.351	5.16
Reinmuth 1	1973.22	7.63	1.995	0.485	5.76
Arend	1967.45	7.76	1.822	0.535	6.02
Oterma	1958.44	7.88	3.388	0.144	4.53
Schaumasse	1960.29	8.18	1.196	0.705	6.92
Jackson-Neujmin	1970.60	8.39	1.428	0.654	6.83
Wolf	1967.66	8.43	2.506	0.395	5.78
Comas Sola	1969.83	8.55	1.769	0.577	6.59
Kearns-Kwee	1972.91	9.01	2.229	0.485	6.43
Swift-Gehrels	1972.66	9.23	1.354	0.692	7.44
Neujmin 3	1972.37	10.57	1.976	0.590	7.66
Gale	1938.46	10.99	1.183	0.761	8.70
Vaisala 1	1971.70	11.28	1.866	0.629	8.19
Slaughter-Burnham	1970.28	11.62	2.543	0.504	7.72
Van Biesbroeck	1966.54	12.41	2.410	0.550	8.31
Wild	1973.50	13.29	1.980	0.647	9.24
Tuttle	1967.24	13.77	1.023	0.822	10.46
Du Toit 1	1974.26	14.97	1.294	0.787	10.85
Schwassmann-Wachmann 1	1974.12	15.03	5.448	0.105	6.73

COMET	Date Last Seen	Period in Years	Perihelion Distance (AU)	Eccentricity	Aphelion Distance
Neujmin 1	1966.94	17.93	1.543	0.775	12.16
Crommelin	1956.82	27.89	0.743	0.919	17.65
Tempel-Tuttle	1965.33	32.91	0.982	0.904	19.56
Stephan-Oterma	1942.96	38.84	1.595	0.861	21.34
Westphal	1913.90	61.86	1.254	0.920	30.03
Olbers	1956.46	69.47	1.179	0.930	32.62
Pons-Brooks	1954.39	70.98	0.774	0.955	33.51
Brorsen-Metcalf	1919.79	71.93	0.485	0.972	34.11
Halley	1910.30	76.09	0.587	0.967	35.33
Herschel-Rigollet	1939.60	154.90	0.748	0.974	56.94
Helfenzrieder	1766.32	4.51	0.403	0.852	5.05
Blanpain	1819.89	5.10	0.892	0.699	5.03
Du Toit 2	1945.30	5.28	1.250	0.588	4.81
Barnard 1	1884.63	5.40	1.280	0.584	4.88
Schwassmann-Wachmann 3	1930.45	5.43	1.011	0.673	5.17
Clark	1973.40	5.52	1.560	0.500	4.69
Brooks 1	1886.43	5.60	1.328	0.579	4.98
Lexell	1770.62	5.60	0.674	0.786	5.63
Kohoutek	1975.04	5.67	1.558	0.510	4.80
Pigott	1783.89	5.89	1.459	0.552	5.06
West-Kohoutek-Ikemura	1975.15	6.11	1.398	0.582	5.29
Kojima	1970.77	6.19	1.632	0.516	5.11
Taylor	1916.08	6.37	1.558	0.546	5.31
Spitaler	1890.82	6.37	1.818	0.471	5.06
Harrington-Wilson	1951.83	6.38	1.665	0.516	5.22
Barnard 3	1892.95	6.52	1.432	0.590	5.55
Churyumov-Gerasimenko	1969.69	6.55	1.285	0.633	5.72
Giacobini	1896.83	6.65	1.455	0.588	5.62
Schorr	1918.75	6.66	1.883	0.468	5.20
Swift	1895.64	7.20	1.298	0.652	6.16
Denning 2	1894.11	7.42	1.147	0.698	6.46
Metcalf	1906.77	7.78	1.631	0.584	6.22
Gehrels 2	1973.92	7.94	2.348	0.410	5.61
Denning 1	1881.70	8.69	0.725	0.828	7.73
Klemola	1965.63	10.97	1.764	0.643	8.11
Boethin	1975.01	11.96	1.096	0.790	9.36
Peters	1846.42	13.38	1.529	0.729	9.74
Gehrels 1	1973.07	14.54	2.935	0.507	8.98
Van Houten	1961.33	15.75	3.939	0.373	8.63
Pons-Gambart	1827.43	63.83	0.807	0.949	31.14
Dubiago	1921.34	67.01	1.116	0.932	31.88
De Vico	1846.18	75.71	0.664	0.963	35.13
Vaisala 2	1942.13	85.42	1.287	0.934	37.50
Swift-Tuttle	1862.64	119.98	0.963	0.960	47.69
Barnard 2	1889.47	145.35	1.105	0.960	54.18
Mellish	1917.27	145.36	0.190	0.993	55.10

matter, the influence of the Earth and Venus, and perhaps even lowly Mercury, would outweigh that of Neptune. So to think that Neptune is somehow responsible for the orbit is to miss the whole point of the matter.

To this point the question of additional planets seems inconclusive, with the balance apparently leaning against the prospect. However, there is one line of reasoning which has not to my knowledge been advanced elsewhere but which I feel is highly suggestive.

The existing model of the Solar System calls for a region of planets extending outward from the sun to Pluto, or roughly 40 a.u. Then we have a blank region until we enter the realm of the comet halo between 30,000 to 50,000 a.u. Being generous, let us postulate a halo doubled in size including all the space between 10,000 to 50,000 a.u. This still leaves a conspicuous gap in the region between 40 to 10,000 a.u., or possibly even between 40 to 30,000 a.u.

To suggest this is all void space would be, I suspect, wholly incorrect. It would be almost impossible to explain such a void by any mechanical means. If we postulate a comet halo, pushing the Solar System out to 50,000 a.u., then we must be prepared to accept responsibility for explaining vast expanses of emptiness if and as they occur. In short, if the Solar system ends at 40 a.u., or 50 or 60 a.u., for that matter, then we are free of the need to explain why the region immediately beyond it is empty. But if we accept the halo, we must also accept the implications of our reasoning and be prepared to talk about the gap between the planets and the halo.

So far as the comet halo is concerned, the evidence for its existence is nearly conclusive. Only two or three comets ever observed had orbits which were hyperbolic, and even these were just barely so. If comets were coming in from outside the system, a clear majority would have hyperbolic orbits, and most of those would be wide hyperbolas, not just marginally so as we find them in our few examples. This is as nearly conclusive as we can hope to get under the circumstances. I know of no contemporary astronomer who seriously doubts the existence of the halo.

This means we must be prepared to discuss the "empty" space between 40 and 10,000 or more a.u.

For my part, I postulate that this region is occupied by literally hundreds of thousands, or even millions of minor asteroids and planetoids possessing radii on the order of 150 to 1,500 kilometers with a few having radii up to roughly 3,000 kilometers and perhaps five or six with radii ranging upwards of 10,000 kilometers. Inclinations and orbits are random in the same sense that comets in the halo appear to possess random inclinations and orbits. There is so much space out there, and motions are so slow, that no systematic scouring has occurred and conditions remain nearly primeval.

Several arguments lead to this hypothesis. A fairly clear line of evidence indicates that planets condense from clouds which contain substantial amounts of particulate matter. A glance at the scarred faces of the moon and Mars is more than adequate to establish this argument and a view of the asteroid belt provides added proof if needed. To suggest that all this particulate matter was confined within a region of some 40 a.u., while simultaneously assuming the comets occupied all space beyond would appear more nearly an article of faith than reason.

Secondly, suppose Lyttleton's hypothesis of the origin of Pluto is correct. If so, this reduces the size of the system to around 30 a.u. and forces the correlary assumption that at this distance there was enough particulate matter to form the nucleus for the condensation of Neptune, Triton, Pluto, and tiny Nereid (radius 120 kilometers) which was obviously a capture from further out. To argue that Nereid was the last such item left over and there is now nothing until we get out to the comet halo requires a truly titanic act of faith on our parts.

Halley's Comet, photographed in 1910 in Santiago, Chile. (Mt. Wilson and Mt. Palomar Observatories)

A third line of reason goes back to Brady and Lowell. If we postulate a very considerable amount of random particulate matter beyond Neptune, then we can arrive at perturbations which give us a vector solution whenever we try to resolve them down to a single object. The discordant mass of Pluto becomes readily understandable as constituting an appreciable fraction of the masses acting on Uranus, but not necessarily the only remaining mass. And Brady's Saturn-sized mass at 58.0 a.u., which is otherwise invisible to telescopes, becomes simply another vector solution. It is a sum of forces rather than an actual object, so naturally there is nothing there to be seen.

A more remote argument comes from the "lost mass" of the galaxy. The physical mechanics of the galaxy require a mass of matter fully 20 percent greater than that we can observe or infer. Such material presumably exists in the form of black bodies: singularities, sub-dwarf stars, planets, free gases, comets, et cetera. We add all this together and still arrive at a shortage of roughly 10 percent. There is just that much mass missing somewhere.

Artist Rick Sternbach depicts Pioneer 11 *flying past ringed planet Saturn, the farthest exploration that our spacecraft have undertaken.*

Conventionally we find our Solar System depicted as consisting of a sun and nine planets plus some miscellaneous objects such as comets, asteroids and satellites. The miscellaneous objects combined would not equal the mass of Earth and the sum of all the dark objects of the system is less than one percent of the mass of the sun. Postulating the existence of the intermediate belt between the inner system and the comet halo changes all this. An aggregate mass several times that of Jupiter could easily exist in this area without being detected, providing it was broken up into enough small fragments. A hundred thousand lunar-sized planetoids would equal 26 Jupiters in mass and would more than adequately account for the "lost mass" of the galaxy, at least so far as our one system is concerned. If this construction were typical of all Solar Systems, the "lost mass" question ceases to be a problem.

So now the matter is turned around. When we began we were talking about the prospects of another planet or two out beyond Pluto. But instead of one or two it turns out the real argument is for the existence of thousands, or hundreds of thousands of planets and asteroids, some of which are in all likelihood approximately the size of Earth.

Styx there is, and Charon too, and stones without number. It's a big, big Solar System!

The Cone Nebula in the region of the constellation Monoceros. Dark clouds of dust, swirling clouds of gas and plasma interact to build new stars and, perhaps, new planetary systems (Mt. Wilson and Mt. Palomar Observatories)

The Origin of the Solar System

Richard C. Hoagland and Ben Bova

Part of the Rosette Nebula, in the constellation Monoceros. The dark shapes are huge clouds of interstellar dust, which, according to theory, form the basic material from which new stars are made. Individual globules of dust are called protostars. (Mt. Wilson and Mt. Palomar Observatories)

We end at the beginning.

The preceding chapters have shown close-up views of five new worlds, together with a new view of our own planet Earth. We have seen that each of these worlds is unique, each very different from every other, a world of its own, intriguing and individual.

Yet for millenia, human thinkers have tacitly assumed that all these different worlds were created at the same time, by the same forces, and probably in a single burst of cosmic creativity.

How can such different worlds have arisen from a single event? Were the planets created all at the same time, in one continuous process? Was the Sun created at the same time, and by the same forces, as the planets and moons and asteroids and the interplanetary dust and gas that pervades our present-day Solar System?

How did it all begin? Over the hundreds of generations of human history, that is the question that theologians, philosophers and scientists have tried to answer. Where did we come from? How did all this come into being?

In many ages and cultures the answer has always been more or less the same. The gods created all that we see: the Sun, the planets, the stars. In fact, in many cases the question was indistinguishable from the answer. The Sun, the Moon, the planets *were* the gods who created this world. But since Galileo's telescope, and the ferment caused by the scientific revolution that is still in progress, modern man has sought real knowledge about the heavens. To paraphrase a thought expressed by Carl Sagan, men of past eras have *wondered;* modern man will *know*.

The difference, in large part, is that we have done something that no past generation of human beings has been able to do. We have left the confines of the Earth and sought our answers amid the reality of the Solar System, with spacecraft such as *Mariner, Pioneer, Zond* and *Venera*. The results of their close-up probes of the planets have filled this book.

Now comes the final—and original—question, seen today in the light of our new knowledge: How did it all begin?

The earliest scientific attempt to answer this ancient query was that of René Descartes, the French philosopher and mathematician. In 1644, Descartes wrote of his conception, in which the primordial material of the Solar System swirled about in vortices, much like the swirls of smoke in a room full of cigar-wielding politicians. Descartes viewed these vortices, these whirlpools and eddies of gas, as gradually cooling and coalescing to form a large central object with smaller solid bodies rotating around it: the Sun and the planets.

About a century later, the German philosopher Immanuel Kant elaborated on Descartes' idea. Still later in the eighteenth century, the French astronomer Pierre Simon de Laplace added his thoughts to the intellectual pot. The result came to be known as the *nebular hypothesis*. Kant and Laplace envisioned the planets forming out of the condensation of a disk-shaped cloud of gas (a nebula) that surrounded the Sun. As the disk-shaped nebula shrank under its own interior gravitational forces (this was the first attempt to bring Newton's law of gravitation into the problem), successive rings of material would be left. These concentric rings of gas would eventually condense still further to form the planets, moons, and other solid bodies of the Solar System.

The nebular hypothesis was simple and elegant, and it survived almost a hundred years, until about 1875. It took one of the giants of science to knock it down: James Clerk Maxwell, the British mathematical physicist who was the prime mover in developing our modern understanding of the behavior of gases.

Maxwell showed mathematically that if the combined masses of all the solid bodies of the Solar System were turned into gas and distributed in a disk-shaped nebula around the Sun, the inner gravitational forces within the disk would not be strong enough to make the gas condense and form solid objects. Quite to the contrary, Maxwell's inexorable mathematics showed that the disk would tend to expand and waft off into space, blowing away like a smoke ring, leaving the Sun quite alone.

The nebular hypothesis was an "evolutionary" type of theory. That is, it pictured the formation of the Solar System as a completely natural sequence of events in the development of a star. Moreover, there seemed no reason why the same kind of event could not happen with other stars, in addition to our Sun. This led to the conclusion that stars acquire planets as a natural part of their evolution. There must be myriads of solar systems in the universe, and many planets where life existed.

But once Maxwell's mathematics turned the nebular hypothesis into a vanishing smoke ring, a completely new kind of theory arose. There were several different variations, but basically they all claimed that the planets and other solid bodies of the Solar System were created long after the Sun came into existence. They saw a chance encounter between the Sun and another star, or a very large comet, pulling material out of the Sun—material that condensed into the planets, et al.

This was a "revolutionary" or "catastrophic" kind of theory. A violent encounter, such as a near-collision between two passing stars, could only happen very rarely because the stars are spaced so far apart. That would mean that our Solar System is a very rare phenomenon, perhaps unique. There is no reason to expect other stars to possess planets, no reason to believe there might be life elsewhere in the universe.

By the turn of the twentieth century, this "catastrophic" theory had many champions, including Thomas Chrowder Chamberlain and Forest Ray Moulton in the U.S., and Sir James Jeans and Sir Harold Jeffreys, two of Britain's most widely-respected astrophysicists.

The Jeans/Jeffreys version of the "catastrophe" was the most popular. They envisioned the Sun being brushed closely by an intruding star. The two came to within several million kilometers of each other—closer than the orbit of Mercury.

To appreciate how rare an occurrence this must be, consider that the average distance separating the stars is on the order of several light-years (10^{13} kilometers) and the average diameter of a star is of the order of 10^6 or 10^7 kilometers. The random velocities of the stars, relative to each other, range around ten kilometers per second. This means that, in the entire 10-billion-year lifetime of the Milky Way galaxy, perhaps only two or three such near-collisions could have happened. Only two or three near misses among more than 100 billion stars! Obviously, if this is the way the Solar System was created, planets are vanishingly rare among the stars.

Jeans and Jeffreys calculated that, as the invading star swept past the Sun, it hauled twin filaments of hot gas from the surface of the Sun. Two filaments: one on the side facing the intruder star, and one on the opposite side of the Sun. This would happen for the same reason that tidal forces create two bulges of high tides in the Earth's seas: one on the side of the Earth facing the Moon, and the other 180 degrees across the globe.

Out of these twin filaments of hot, dense gas (not the thin, cold gas of the old nebular hypothesis) Jeans and Jeffreys claimed that the planets condensed.

Early in the twentieth century this was a compelling account of the formation of the Solar System. On the basis of pure dynamics, the exchange of material and velocities seemed to explain adequately most of the observed features of the Solar System: the essential "flatness" of the orbital planes of the planets, the fact that they all orbited in the same west-to-east direction, the near-circularity of their orbits. Of course, if Jeans and Jeffreys were right, then amid the vast star clouds of the Milky Way, across eons of galactic time, there could be only one other planetary system, orbiting that unknown intruder star, lost forever amid the glow of totally barren suns.

It was a stark picture, appropriate to the Victorian legacy of the uniqueness of man. It meant that life, particularly intelligent life, must be unique to our world.

Of course, it was an incorrect picture.

Henry Norris Russell, the American astronomer, pointed to the major flaw in the "catastrophic" theory. The Sun is rotating too slowly.

If the stellar-encounter origin of the Solar System were true, the Sun should have most of the *angular momentum* of the Solar System. It should be spinning fast, something like 500 kilometers per second or more. But the Sun actually spins quite slowly, more like two kilometers per second. It revolves once in about a month.

The Great Nebula in Orion, a breeding ground for new stars. (Mt. Wilson and Mt. Palomar Observatories)

Russell's mathematics showed that if an intruding star had sideswiped the Sun, the result would be a fast-spinning Sun, not the slow spin we actually observe. In fact, 98 percent of the angular momentum of the Solar System resides in the orbital and rotational motions of the planets; only two percent resides in the rotation of the Sun—which contains a thousand times the mass of all the planets, combined.

Moreover, when Maxwell's equations were re-examined, they showed that a hot gaseous filament pulled from the Sun's surface or interior would also dissipate into space, much as the old cold gas of the nebular hypothesis. The hot gas might be denser, but its higher temperature would tend to make it dissipate faster than its density would tend to make it coalesce.

Back to Square One.

By the middle of the 1930s all the theories about the Solar System's origins were in a shambles. As one frustrated astronomer fumed, "The damned thing must have had some kind of origin, because it's here!" Scant consolation.

There have been many times in the history of science when understanding has been delayed, simply because the human race had not yet discovered or invented the kinds of tools needed to reach that new understanding. Copernicus' theory could not be confirmed until telescopes came into astronomy. Newton needed calculus to form his laws of motion and gravitation, so he invented it. Astronomers seeking to understand the origin of our Solar System needed other tools. As we have already seen, one of the new tools has been the spacecraft probe that brings close-up information about the planets back to us. But even before that, a new understanding of the basic nature of matter was necessary for those who sought the origins of the Solar System.

Astronomers and astrophysicists had to learn about *plasmas*.

Plasma has been called the fourth state of matter: solids, liquids, gases, *plasmas*. Actually, in terms of cosmic abundance, plasma is the first state of matter. Most of the universe is plasma. The stars, including the Sun, are plasma. Much of the thin clouds of matter wafting between the stars is plasma, rather than gas. Only in cold, out-of-the-way corners of the universe such as minor planets like Earth do we find solids, liquids, and gases predominating over plasmas.

A plasma is an ionized gas. That is, some (or all) of the atoms in the gas have been stripped of some of their orbital electrons. In a normal gas, the atoms are intact and electrically neutral.

But in a plasma, enough energy has been applied to the atoms to wrench some of their orbital electrons away from them. Thus we have a mixture consisting of some free electrons, which carry negative electrical charges; some bereft ions (atoms that have lost electrons, and are therefore positively-charged); and some neutral atoms. In a completely-ionized plasma, there are no neutral atoms remaining.

For every free electron in the plasma, there's a positively-charged ion, so that on the whole the plasma is electrically neutral; it contains as many negative charges as positive. But those electrons are free and very mobile; they can carry electrical currents and be influenced by magnetic fields. A plasma, *unlike a gas,* can be shaped, moved, heated, influenced by electromagnetic forces.

Stars are totally plasma; our Sun included. Planets are essentially composed of the less-energetic forms of matter: solids, liquids and gases. Plasma occurs naturally on Earth only around lightning discharges. The insides of fluorescent lamps are highly-ionized plasmas; the exhausts of rocket engines are very slightly-ionized plasmas.

Plasmas were discovered early in the twentieth century, but it wasn't until after World War II that physicists began studying the dynamics of plasmas—how plasmas behave when they are in motion. A new field of investigation arose, and earned the jawbreaking name of *magnetohydrodynamics* (MHD): the study of the interactions between electromagnetic forces and plasmas.

The Swedish physicist Hannes Alfvén applied plasma physics and magnetohydrodynamics to the problems of the Solar System's origin, and earned the Nobel Prize in 1970 for his work.

Alfvén returned to the old nebular hypothesis of Kant and Laplace and showed that if the original solar nebula—the cloud out of which both the Sun and planets were formed—had been a plasma, then many of the problems that had plagued the earlier theories could be cleared away.

This new approach, impossible until the intricate interactions of plasmas and magnetic fields could be understood, was the breakthrough needed to begin a workable theory of the Solar System's origin. Moreover, this new "plasma-nebular hypothesis" links up with observational evidence from other stars in the heavens, and leads to the conclusion that planetary systems are a natural, evolutionary part of the life histories of many stars.

Astronomical observations and theoretical developments point toward the conclusion that stars are created out of clouds of gas and dust that pervade interstellar space. Such swirling clouds as the Orion Nebula are now known to be breeding grounds for new stars. These gas and dust clouds make up about two percent of the total mass of the Milky Way galaxy; the bulk of the galaxy's mass is tied up in its 100 billion-plus stars.

About 99 percent of the universe's matter consists of hydrogen, the simplest and lightest of all the elements. No matter what state the matter is in—solid, liquid, gas or plasma—almost all of it is hydrogen. Elements heavier than hydrogen, from helium on up to the transuranium elements, make up the remaining one percent of the universe's atoms. And these heavier elements get progressively rarer as they get heavier and their atoms become more complex.

Calculations, backed by optical and radio observations, indicate that single stars cannot condense out of the thin "reservoir" of interstellar gases all by themselves. It takes a mass of material equal to about a thousand times the mass of the Sun to initiate the condensation processes that lead to the birth of a new star. The observational evidence shows that new stars are typically formed in vast clouds of swirling interstellar gas, dust and plasma, clouds that have total masses of several million suns. This is where the gravitational collapse that leads to the creation of new stars can begin.

Thus, in the beginning, our Sun was probably born in a cluster of other stars, all gestated from a massive cloud of interstellar material—predominantly hydrogen. The cloud collapsed due to gravitational forces within itself. At a certain point in this gravitational collapse, portions of the cloud became dense enough for individual stars to start forming.

This was between four-and-a-half and five billion years ago. Our Sun was a huge amorphous cloud of dust grains and very slightly ionized hydrogen, with a smattering of heavier elements. The cloud bore a very weak magnetic field. It was essentially spherical in shape, with a diameter on the order of a light year (10^{13} km), or ten million times its present size.

The dust grains were merely microscopic flecks of frozen hydrogen, no more than a hundred-thousandth of a centimeter (10^{-5} cm) long, with a few atoms of heavier elements mixed in with the hydrogen.

At this point, as the cloud slowly rotated and sank inward on itself, it developed turbulent eddies and currents that caused random variations in its density. These "lumps" became the comets. Most theorists agree that the Solar System is ringed with a "halo" of comets orbiting in the cold and darkness far beyond Pluto's orbit. Occasionally a gravitational perturbation sends a member of their family plunging in toward the Sun, sometimes creating a spectacular show in the skies of Earth.

These comets are little more than frozen iceballs of primordial gases; they represent samples of the original material of the solar system, still in its original form. Some astrophysicists have suggested sending probes to Halley's Comet, when it returns in 1985, to gather samples of the Solar System's original ingredients.

The collapse of the solar cloud continued, some four and a half billion years ago. This condensation phase happened in an astronomical twinkling: from the first fragmentation of the solar cloud to the moment when the Sun lit its internal nuclear fires and turned into a full-fledged star probably took no more than ten million years; certainly not more than 100 million. Compared to the 10-billion-year stable lifetime of our Sun, and the 100-billion-year "old age" phase that it will eventually retire to, its birth was dramatically, incredibly swift.

As the cloud sank in on itself and condensed, it spun faster and faster. Rotational energy (angular momentum) cannot be created or destroyed within a closed system. Like any form of energy, it is *conserved*. Take a huge, slowly spinning cloud and shrink it; it will spin faster and faster as it gets smaller.

And the weak magnetic field that permeates the light-year-wide cloud will become progressively stronger as it, too, condenses. The same amount of total magnetic energy is being squeezed into a constantly-smaller volume of space. Magnetic field intensity increases, until at last the magnetic field can begin to have a discernible effect on the motions of the ionized gas cloud.

The conservation of angular momentum was forcing the solar cloud to spin faster and faster. If no other forces had intervened, the final equatorial velocity of the cloud, once the collapse had stopped, would have been something over 1,000 kilometers per second. If the Sun truly spun that fast, it would have dissipated as it collapsed, growing a thin equatorial gaseous ring by the time it had shrunk to the size of Mercury's orbit. The remaining star, after whirling this gaseous disk away from itself, would end up alone and spinning at several hundred kilometers per second.

But angular momentum was not the only force present, and the solar cloud was not a gas; it was a plasma.

Electrically charged objects, be they ions, electrons, or spacecraft, are inhibited by a magnetic field from crossing the field's lines of force. Motion *along* the lines is not prevented. For large, massive objects such as spacecraft, the hindering effect of the magnetic field is unnoticeable. The spacecraft wins because it has too much mass and energy of motion for the field to deflect it. For a single ion or electron, the situation is very different. If the magnetic field is above a certain intensity, the ion or electron must follow where the field guides. If the field is weaker than this critical level and the energy of motion of the particle is greater than the field's energy, then the particle wins, and the field is dragged along by the ion or electron.

For a great mass of collapsing gas that is only slightly ionized, the gas would win over the magnetic field, calculations show. The collapsing cloud would produce a magnetic field much like the spokes of a huge bicycle wheel, but it would be a weak field, somewhat like spokes made of boiled spaghetti. However, as the collapse continues and the density of the plasma rises, the magnetic field intensity increases. The field lines get stiffer. The inner parts of the cloud tend to collapse faster than the outer fringes (remember, it's still almost a light-year across!) and the inner portions will rotate around the center of mass faster than the outer layers. A shrinking, egg-shaped nebula results, collapsing at right angles to the plane of rotation, unhindered by such complicating factors as centrifugal force, particle collisions, and energy dissipation in the form of heat.

As the collapse proceeds, the magnetic field in the core of the cloud gets wrapped around the center of the cloud by the faster rotation of the inner parts of the revolving nebula. This wrapping enhances the intensity of the field, and eventually results in the acceleration of particles—ions and electrons—from the conversion of magnetic rotational energy. Enormous superflares begin to blast out from the heart of the nebula, the central region that is on its way to becoming a star, our Sun. These titanic flares release huge amounts of energy inside the nebula, which serves to ionize more of the gas, make it more of a plasma, increase the electrical conductivity of the nebula, and convert rotational energy into radiation, charged particles, and heat.

In other words: all hell breaks loose.

This electromagnetic heating, adding to the gravitational heating from the collapse, serves to ionize a large percentage of the nebula. Now, if we envision the stretched magnetic field, much like the spokes of a wheel stretched around the hub of the wheel by slippage of the rim relative to the hub, we can see how the faster rotation of the now "stiff" field can sweep the outer parts of the highly-flattened, spinning disk along.

Angular momentum is transferred from the central core via this now-strong magnetic field to the mass of the plasma in the flattened disk. The rotation of the central proto-Sun is slowed and the rotation of the disk or plasma surrounding it is accelerated. Through millions of years the magnetic field acts to transfer this energy of rotation, spinning up the exterior disk by removing rotational energy from the central core. At the same time, the proto-Sun is alive with incredible explosions, superflares releasing as much energy in a few seconds as the modern Sun emits in hundreds of hours.

Thus the angular momentum of our present Sun was reduced and the magnetic energy originally contained in the solar nebula was dissipated, converted into the energy of accelerated particles and energy transfer to the exterior parts of the nebular disk. A mass of plasma and dust some 100 times greater than the present mass of the Sun collapsed from the interstellar medium, spun up, formed a disk-shaped nebular plane around the massive central body, underwent extensive electromagnetic heating, converted rotational energy into energetic particles, and transferred a major amount of material into progressively further orbits around the central proto-Sun, leaving the not-yet-shining star rotating sedately once a month rather than once every few hours.

But we still do not have planets.

This abbreviated history of the birthing star system is a combination of theory and observation. We've given you the theory. Now let us take a quick look at some key observations that support the theory and make us believe that we've just described reality, not fiction.

First: On the matter of angular momentum. The late Otto Struve, former director of the National Radio Astronomy Observatory in Green Bank, West Virginia, devoted much of his life to studying the rotations of stars. He discovered a very revealing statistic about stellar spin rates. There is a definite tendency for massive, bright, high-temperature stars—which are relatively young—to rotate rapidly, several hundred kilometers per second. Stars like the Sun—yellowish, 6000°K or cooler, moderate mass, several billion years old—*invariably* rotate quite slowly. Where the angular momentum of stars is concerned, there is a sharp breakpoint: stars about 2000° hotter and 1.5 times the mass of the Sun spin fast. Older, cooler, smaller stars spin slowly.

Struve concluded that the older stars rotate slowly because they have transferred their angular momentum to planets, or a nebular cloud, rotating around them. The younger stars rotate fast because they either have no cloud to transfer the momentum to, or they have not had enough time for the transfer to take effect.

Second: On the matter of super-flares. Astronomers have found a certain type of star that gives observational proof of our theory of stellar formation. They are called T Tauri stars, after the first of their type to be investigated, the star known by the catalogue designation T Tauri, in the constellation of Taurus, the Bull.

T Tauri stars are found embedded deep within a nebula, and they are always highly unstable, flaring and flickering at an apparently random rate. They are always accompanied by a mass of plasma expanding outward at relatively rapid rates. If you were outside the proto-solar system in the last stages of the Sun's formation, what would you see? Something very like a T Tauri star. The plasma and dust cloud is naturally the birthplace of the star and its companions, which fragmented out as the parent nebula condensed from the original cloud of several thousand solar masses.

The flickering and flare-like luminosity is also easy to explain. To an outside observer the final derotation and momentum transfer stage of the proto-Sun would have all the observable features of a randomly variable star; the unpredictable superflares would cause enormous brief increases in the star's light output. T Tauri stars fit beautifully the particulars for what the young Sun would have looked like at the birth of the Solar System.

To resume...

When we left our still-forming Solar System, the proto-Sun had successfully transformed most of its angular momentum to the nebula surrounding it, via the compressed magnetic field. But there were still no planets in existence.

Somewhere between the violence of the T Tauri stage and the quietude of the present, the planets were formed. It has taken the application of almost every branch of science, including exploration of the Moon by the Apollo astronauts, to begin to puzzle together just how this happened.

In the past thirty years, scientists from many different specialized disciplines have made their contributions to astrophysics. Carl F. Von Weizsacker and Gerald P. Kuiper (physicist and astronomer, respectively) used sophisticated concepts of fluid dynamics to help attack the problem, in the 1940's. Harold C. Urey brought chemistry into the picture in the 1950's. Alfvén and Fred Hoyle, also in the 1950's, utilized plasma physics.

It has become apparent that no simplistic theory—be it gravitational, magnetic, chemical, or whatnot—can by itself explain the rich diversity of the Solar System. Explaining the formation of the planets is probably the most complex problem of all astronomical history.

But a combination of these specialized areas of research, an application of essentially all the knowledge science has amassed about the behavior of matter, is necessary if we are to penetrate the fascinating mystery that produced the Earth and all the other solid bodies of our Solar System.

The following, then, is a synthesis: Thirty years of astrophysics applied to understanding the complexity of forces that shaped our Solar System. And what we know today is only the rough outline of the whole story.

It now seems apparent that the formation of the Solar System occurred in two main phases, probably with some overlap between them. Phase I was the collapse of the original cloud and the production of a slowly-rotating central body. At some point in this collapse, the temperature and pressure at the heart of the proto-Sun reached the level where nuclear reactions are ignited. The proto-Sun turned into a true star: it began to shine with the energy of *thermonuclear fusion,* an energy source that will keep the Sun shining steadily for ten billion years, at least.

In thermonuclear fusion, nuclei of the lightest element, hydrogen, are welded together to produce nuclei of helium, plus energy. Although the actual reactions are more complex than this simple explanation, it is the energy of thermonuclear fusion that powers the Sun and the stars.

As its T Tauri phase was ending, the Sun settled into what is termed its Hayashi Phase (named after the astrophysicist Chushiro Hayashi who calculated this aspect of solar evolution). This brief period saw the Sun shining at 100 times its present brightness. It also still blasted out very energetic solar flares, much stronger than the flares the Sun emits today.

Phase II, the creation of the planets, must have occurred around this brilliant, over-luminous Sun. It was still surrounded by a disk of dense plasma. This nebula, highly heated near the center, was naturally cooler and less dense out toward its farthest regions. It was within this heated, swirling nebula that the planets came into being.

Before the Space Age and the landings on the Moon, the only extraterrestrial samples of material that scientists had access to were meteorites, the remains of "shooting stars" that had fallen from interplanetary space. There are two major varieties of meteorite: *stones* and *irons*. Stony meteorites far outnumber the irons. Their structure provided early clues to the formation of the meteorites themselves, as well as the larger bodies of the Solar System.

Embedded in most of the stones are tiny, spherical inclusions called *chondrules.* Primarily composed of two minerals—olivine and proxene—these tiny droplets permeate the stony structure which is itself composed of fragmented crystals of the same materials. Loosely called *silicates* because the element silicon is present (together with magnesium, iron and oxygen), these minerals are thought to make up a large percentage of the mass of the planets closest to the Sun, the so-called terrestrial planets—Mercury, Venus, Earth, the Moon, and Mars.

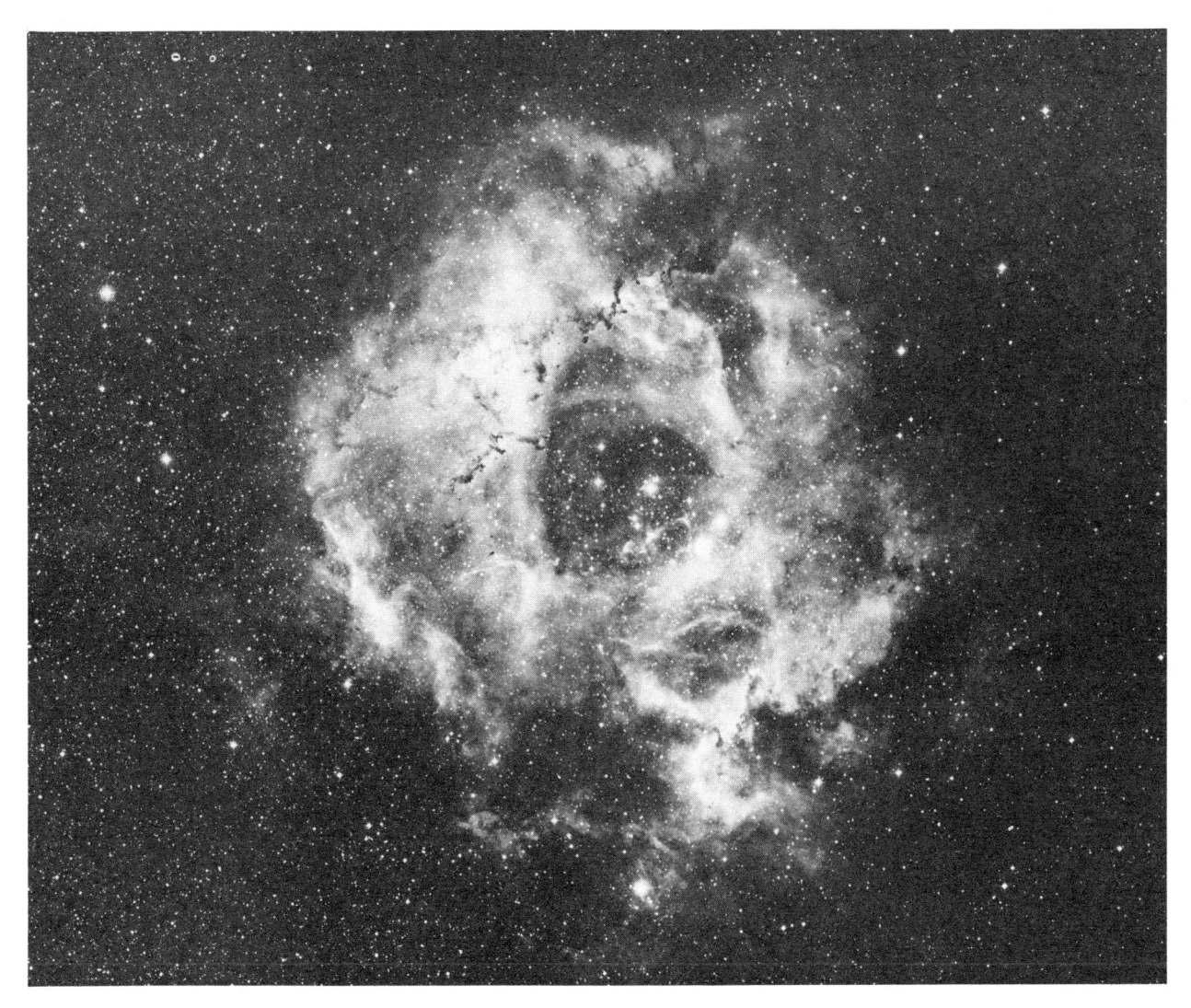

The Rosette Nebula in its entirety. Hot young stars in the center of the nebulosity have literally blown the glowing gas and dust away from themselves with the intensity of their energetic radiation, thereby creating the "doughnut" effect. (Mt. Wilson and Mt. Palomar Observatories)

The chondrules were apparently formed under weightless conditions in interplanetary space, and represent the basic building blocks of the meteorites, and perhaps of the planets as well. In any body of moderate size, the chondrules will tend to be crushed out of shape and become unrecognizable. Thus the stony meteorites preserve a portion of the early history of planet-building.

In the original solar nebula, dust-sized grains of solid material were sprinkled through the slightly-ionized plasma. As the nebula heated up, however, these grains were vaporized and the cloud became entirely gaseous and highly ionized. As the Sun began to settle down from its early explosive behavior, the nebular disk radiated away much of its heat, dissipating it into space. A flattened disk has plenty of surface area and makes an excellent radiator.

As the disk cooled, solid grains could once again condense out of the plasma. Now, however, they were most likely spherical in form, solidified droplets of matter, chondrules whose chemical composition depended very heavily on the temperature of the disk at the particular site where they condensed.

The density and temperature within the disk varied outward from the Sun to such a degree that condensation of different minerals was favored at different distances from the Sun. Nearest the Sun, where the disk temperatures were in the neighborhood of 1600°K, only the most refractory elements and minerals could condense, such as oxides of calcium and aluminum, CaO and Al_2O_3 Familiar elements such as silicon, carbon and even most metals would remain vaporized at that temperature. Farther out, in the vicinity of Earth, the temperature permitted a rain of silicates and iron compounds. At greater distances and lower temperatures, water and ammonia could condense, and even further, methane and solids of such elements as argon.

Elements, minerals, oxides, hydrates, and solids thus rained out of the nebula at specific distances and temperatures, according to their chemical properties.

The result: The gradient of planetary masses, densities, and compositions that we observe today. The inner planets of the solar system are small and dense, because they were created out of the high-temperature materials that could condense so close to the Sun's warmth. The lightest elements, hydrogen and helium, accounted for more than 99 percent of the atoms in the nebula. The heavier elements—carbon, oxygen, nitrogen, neon, argon—together account for less than one percent.

So small, low-mass planets composed of silicates and metals were formed nearest the Sun: Mercury, Venus, Earth and Mars. The variation of each planet's individual composition and density is very highly dependent on the temperature of the nebula at that planet's distance from the Sun. Since the planets formed, they have differentiated, or separated into heavy cores and lighter crusts, and many of the heavier minerals have sunk far below the surface of the Earth. Thus our newly won knowledge of the chemistry of the Moon, and the spaceprobe evidence for volcanism on Mars and the lack of it on Mercury are our only clues as yet to the detailed chemistry of these other so-called "terrestrial" worlds. But basically the premise holds: the terrestrial group of planets are small and dense because of the low abundances of the silicates and metals that could condense at such close distance to the Sun.

Billions of condensing droplets, each orbiting around the seething heart of the flaring Sun, solidifying, colliding with other chondrules, melding, growing as the larger bodies swallow up the smaller. A sweeping accretion of smaller particles into larger bodies builds up the nuclei for the growth of worlds.

Although far less abundant than hydrogen and helium, oxygen exists in large enough amounts to be a major element in the formation of the planetary system. It is found in oxides, the various complex minerals of meteorites, in the rocks of the Moon, and is present even on Mars, as "hydrogen monoxide:" water.

Somewhere in the vicinity of Earth oxygen combined chemically with hydrogen to form water, and thus was trapped, along with the silicates and metals that formed our world.

Beyond the orbit of Mars, across the falling curve of temperature and pressure in the nebular disk, the prevalent water vapor turned into droplets. And slightly farther out, directly into snow. Imagine, if you can, a vast blizzard of frozen particles, snowflakes whirling for eons like an interplanetary storm in a vast torus around the Sun. Three-quarters of a billion kilometers from the shimmering intensity of the Sun, this enormous ring of winter collects itself, a swirling tube of white, of spinning shards of ice, of slushes coating growing nuclei of icebergs adrift in space.

Far more abundant than the other compounds that have rained from the nebula closer to the Sun, the snowstorm produces quickly-growing bodies that will dominate the outer planetary system. With these frozen balls of water, ammonia, and even silicates to act as centers of gravitational attraction, the still-gaseous elements within the cloud can fall and stay on these growing proto-planets. Miniature nebulas develop, twin whirlpools of swirling snow slushes, hydrogen and helium stolen from the major solar disk: thus the growing cores of Jupiter and Saturn eventually became the most massive members of the Sun's family. Their masses, densities, and chemical compositions reflect the highly favorable combination of environment and elements out of which they were composed.

A bit more material, and Jupiter would have become a star. The fact that double and multiple star systems appear to account for more than half of all the known stars in the sky would indicate that planetary systems and multiple star systems are quite closely related.

Beyond the planets Jupiter and Saturn (which orbits a frigid billion and a half kilometers from the Sun) the nebula dramatically decreased in temperature and density, producing worlds apparently composed mainly of methane, ammonia and neon, with traces of other elements. So little is known about the chemistry of Uranus and Neptune that all that can be said about these massive worlds (each is about 15 times Earth's mass) is that they appear to be vastly deficient in the most abundant elements, hydrogen and helium. At present there is no satisfactory explanation of this anomaly.

Of Pluto, little can be said except that Lyttleton's hypothesis—that it is an escaped satellite of Neptune—seems best to fit its appearance. Pluto is a small, terrestrial-sized world rotating once every six days, which is far more slowly than any of the gas giants.

This, then, is the best current thinking about the origin of the Solar System. There is an enormous amount that still lacks satisfactory explanation. The system is filled with maddening anomalies. For example: how did Mercury get its three-halves rotation? Why does Venus rotate in retrograde, and how did its rotation come to be locked in resonance with Earth? What about the double-planet system of Earth and Moon? And their vastly different densities?

A startling anomaly has come out of the space probe close-ups of the new worlds: All the terrestrial planets, including the Moon, have strong hemispherical assymetry. Each planet, from Mercury to Mars, has one hemisphere that is more or less featureless and the other crammed with the major morphological feature inherent to that world, be it craters, continents, or maria.

No one can yet account for the smooth decrease in density of Jupiter's four major Galilean satellites, from Io's 3.5 to Callisto's 1.7. Nor has anyone yet produced a satisfactory explanation for Saturn's rings. They cannot be just water compounds, because they give radar reflections that are much too strong for water ices!

The details of the Solar System that have *not* been explained could fill another book: a book of blank pages.

But then, we have really only recently discovered how to write.

We stand poised on the threshold of incredible discoveries. Those of us who have lived through this breathtaking wave of revolution that has swept from Mercury to Jupiter can remember the barren, sterile atmosphere of theory surrounding the origin of the Solar System before the advent of the Space Age. But now, with man-made robots spreading across the planetary system in a wave of new discovery, with each planet-fall signifying the end of many old theories and the rise of new ones, the field of planetology is thriving as it never has before.

...AND BEYOND

We have spread our quest for knowledge to the farthest reaches of the Solar System. Beyond lie the stars, millions of times farther away than the most distant planet.

What lies beyond today's quest? Beyond, not in terms of distance, but in terms of *time*.

The picture presented by our future-looking writers is an exciting one. Self-sufficient bases on the Moon, factories orbiting the Earth and utilizing raw materials from other worlds, scientific teams living on Mercury and Mars, re-forming the cauldron of Venus into something more Earthlike, communicating with the living intelligent denizens of Jupiter.

Perhaps these are all nothing more than sciene-fiction pipedreams. But when I was in school, so was reaching for the Moon a science-fiction pipedream. So were lasers and artificial hearts and nuclear powered ships.

We have seen new worlds closeup. They look bleak and forbidding when compared to our Earth—and yet they are terribly exciting and compelling. We will send more robot probes to these worlds. And then teams of scientific explorers. And someday, inevitably, human colonists will set sail for these harsh new worlds, for as many and varied reasons as colonists once turned their faces westward, toward the New World of the Americas.

It will not happen in this century, perhaps. But it has already begun.

Just as *Pioneer 11* is now coasting toward the planet Saturn obedient to the universal forces of gravity and momentum, so will human beings respond to the forces that shape their lives and their societies. We have sent our scouts to five new worlds. We have set foot on the Moon. Our future will forever encompass the new worlds of the Solar System. They are there in space, waiting for us. And we will answer their call.

Glossary

Albedo The fraction of incident sunlight that a planet, satellite or asteroid reflects.

ALSEP Acronym for Apollo Lunar Surface Experiment Package, a collection of scientific instruments taken to the moon and left on the surface by the *Apollo* astronauts.

Angular Diameter The angle subtended by the diameter of an object, expressed in degrees, minutes and seconds of arc. The angular diameter of the moon is about half a degree, or 30 minutes.

Aphelion The point in an orbit of an object around the sun where the object is farthest from the sun; opposite of perihelion.

Asteroid Belt The doughnut-shaped zone around the sun between the orbits of Mars and Jupiter, extending from 2.1 to 3.5 astronomical units, in which are found thousands of asteroids (also called minor planets or planetoids) of irregular shapes, having diameters that range from a fraction of a kilometer to about 780 kilometers.

Astronomical Unit A measure of distance in the Solar System, equal to the mean distance between the earth and the sun; its value is approximately 149 million kilometers.

Atmosphere (Pressure) The average pressure at sea level on the Earth, equal to 1.003 kilograms per square centimeter. (14.7 pounds per square inch)

Attitude A spacecraft's orientation in space. It can be changed, usually by firing jets of gas, so that solar panels and other instruments can be pointed at target bodies.

Aurora The light radiated by atoms and ions in the ionosphere of the Earth, usually at the north or south pole.

Bands (in Spectra) Emission or absorption lines, usually in the spectra of chemical compounds, that are so numerous and closely spaced that they coalesce into broad emission or absorption bands.

Bit A unit of information equivalent to "on" and "off" or "1" and "0" in binary code.

Bode's Law A sequence of numbers that gives the approximate distances of the planets from the Sun in astronomical units.

Bow Shock Wave The interface formed where the electrically charged solar wind encounters an obstacle in space such as the atmosphere or the magnetic field of a planet. Behind the shock wave the speed of the solar wind is less than the speed of sound.

Caldera A large enclosed or partially enclosed depression caused by the collapse or explosion of a volcano.

Celsius A centigrade temperature scale in which water freezes at zero degrees and boils at 100 degrees; denoted °C.

Convection Current A column of hot material rising through cooler material because of their difference in density.

Deuterium A "heavy" isotope of hydrogen in which the nucleus of the atom

219

Doppler Shift
consists of one proton and one neutron instead of only one proton.
The apparent change in the wavelength (or frequency) of the radiation from a source due to its relative motion in the line of sight. If the source is approaching the observer, its radiation is shifted toward the blue, or shorter, wavelengths of the spectrum; if it is receding, its radiation is shifted toward the red, or longer wavelengths.

Ecliptic
The plane in the Solar System defined by the Earth's orbit around the Sun; also, the apparent annual path of the sun from west to east around the sky as seen from the earth, which results from the revolution of the Earth around the Sun.

Eccentricity
Loosely, how elliptical the orbit of a body is; the eccentricity of an orbit is defined as the ratio of the distance between the two foci of the ellipse to the length of the ellipse's major axis. An orbit with an eccentricity of 0 is circular; of between 0 and 1 is elliptical; of 1 is parabolic; greater than 1 is hyperbolic.

Electromagnetic Radiation
Radiation consisting of waves propagating through the building up and breaking down of electric and magnetic fields; the electromagnetic spectrum includes radio, microwave, infrared, visible light, ultraviolet, X-rays and gamma rays.

Erosion
A group of natural processes by which rock and soil material is removed from one place on a planetary surface; it includes weathering, dissolution, abrasion, corrosion and transport of material by wind.

Fault
A break in the continuity of a rock form caused by shifting or dislodging of the planetary crust.

Flyby
A space mission in which an instrumented vehicle passes a planet without going into orbit, entering the atmosphere or landing on the surface; examples are the *Pioneer* missions to Jupiter and the *Mariner* missions past Mars, Venus and Mercury.

Frequency
The number of cycles of an electromagnetic wave that passes a point in space in one second. The frequency, ν, of the wave is related to its wavelength, λ, by the formula $\nu = c/\lambda$, where c is the speed of light (300,000 kilometers per second). The unit of frequency is the hertz, or cycle per second.

Imaging Photopolarimeter
A sensor for measuring the brightness and the polarization of light; usually mounted to scan the target so that the readings can be assembled into a picture.

Inclination
The angle between the orbital plane of a revolving body and some fundamental plane—usually the plane of the Ecliptic; also the angle between the body's equatorial plane and the plane of its orbit.

Infrared
Electromagnetic radiation having wavelengths longer than visible light but shorter than microwaves, and having a penetrating heating effect in planetary atmospheres.

Infrared Radiometer
Instrument for measuring the temperature of an object from the intensity of its radiated heat.

Ion
An atom that has become electrically charged by the addition or loss of one or more electrons.

Ionization
The process by which an atom gains or loses electrons.

Ionosphere
Upper region of a planet's atmosphere in which many of the atoms are ionized. On the Earth the ionosphere extends from about 50 kilometers to 400 kilometers and is caused by the ionization of rarefied atmospheric gases by solar energy.

Kelvin
A centigrade temperature scale in which zero degrees is at

	absolute zero (-273°C) where all molecular motion stops, where water freezes at +273 degrees and boils at +373 degrees; designated °K.
Lagrangian Points	Five points in the plane of revolution of two bodies revolving mutually around each other in circular orbits, where a third body of negligible mass can remain in equilbrium with respect to the other two bodies.
Lander	Spacecraft or mission that lands on another celestial body, such as *Surveyor* on the Moon or *Viking* on Mars.
Light-Year	The distance light travels in a vacuum in a year, equal to 9.46 x 10^{12} kilometers.
Limb	Apparent edge of a celestial body as seen in the sky.
Limb Darkening	The phenomenon by which the atmosphere of a body is darker near the limb than near the center of its disk.
Magma	Molten rock in the interior of the earth surrounding the core and underlying the crust.
Mare	Latin for "sea"—name applied to many of the dark flat "sea-like" features on the Moon, Mars and Mercury. The plural is MARIA.
Mascon	An area of mass concentration or high density within a planetary body, usually near the surface.
Occultation	An eclipse of a star or planet by the moon or another planet; also the eclipse of a planet's satellites or of a spacecraft by the planet.
Orbiter	Spacecraft or mission involving the insertion of a vehicle into orbit around another celestial body.
Organic	Designating a chemical compound containing long chains of carbon-based molecules.
Oblate Spheroid	A spheroid having an equatorial diameter greater than the polar diameter.
Perihelion	Point in an orbit of an object around the sun where the object is closest to the sun; opposite of aphelion.
Plasma	An ionized gas containing approximately equal numbers of free ions and electrons, as in the solar wind.
Plate Tectonics	The theory of the formation of continents and oceans due to the movement of crustal material propelled by convection currents in the underlying magma.
Radar Astronomy	Study of the motion and form of other planets with radar equipment; it differs from radio astronomy in that signals are transmitted by the observer and reflected by the object of interest.
Ray (lunar)	Any of a system of bright elongated streaks, sometimes associated with a crater on the moon.
Resolution	The degree to which fine details in an image are separated or resolved.
Rill	Sinuous crevasse in the Moon's surface.
Scarp	A line of steep cliffs produced by faulting or erosion.
Seeing	The unsteadiness of the earth's atmosphere that blurs images of stars and planets.
Shield Volcano	The largest type of volcanic cone, which resembles a shield or low, sloping dome in profile.
Solar Flare	A sudden and temporary outburst of light and particles from an extended region on the surface of the Sun.
Solar Wind	Plasma blown constantly at supersonic speed out of the Sun in all directions; it consists of electrons, protons, alpha particles (helium nuclei) and some heavier ions.
Sublimation	A substance's change of state directly from the solid to the gaseous phase, such as with carbon dioxide.
Sunspot	A temporarily cool and relatively dark region on the surface of the sun associated with strong solar magnetic fields.
Tidal Force	A differential gravitational force that tends to deform a body.

Tide — Deformation of a body by the differential gravitational force exerted on it by another body; in the Earth, the deformation of the ocean (and to a lesser extent, of the land and air) by the gravitational pull of the Moon and Sun.

Torr — A unit of pressure, equal to 1/760th of an atmosphere.

Ultraviolet — Electromagnetic radiation of wavelengths shorter than the shortest visible wavelengths (violet) and longer than soft X-rays.

Ultraviolet Spectrometer — An optical instrument for analyzing the intensity of ultraviolet light at various wavelengths.

Van Allen Belts — Two doughnut-shaped regions surrounding a planet where many rapidly moving charged particles are trapped in the planet's magnetic field.

Wavelength — The spacing of the crests or troughs in a wave train. The wavelength λ of electromagnetic radiation is related to the frequency ν by the formula $\lambda = c\nu$, where c is the speed of light (300,000 kilometers per second). Wavelengths of light are generally expressed in Angstroms (one Å = 10^{-10} meter) or nanometers (one nm = 10^{-9} meter); wavelengths of radio waves are expressed in centimeters and meters.